FIRST
AID IN
MATHEMATICS

Robert Sulley

HODDER
EDUCATION
AN HACHETTE UK COMPANY

Acknowledgements

With thanks to David and Sally for their helpful comments on the manuscript, to Phillipa, Gina and Laurice for copy editing the manuscript and proofreading the text, and to Fiona and Rosa for their support during the writing of this book.

Hachette UK's policy is to use papers that are natural, renewable and recyclable products and made from wood grown in sustainable forests. The logging and manufacturing processes are expected to conform to the environmental regulations of the country of origin.

Orders: please contact Bookpoint Ltd, 130 Milton Park, Abingdon, Oxon OX14 4SB. Telephone: (44) 01235 827720. Fax: (44) 01235 400454. Lines are open 9.00–5.00, Monday to Saturday, with a 24-hour message answering service. Visit our website at www.hoddereducation.co.uk

© Robert Sulley 2010
First published in 2010 by
Hodder Education, an Hachette UK Company,
338 Euston Road
London NW1 3BH

Colour edition first published 2014

Impression number   10 9 8 7 6 5 4 3 2 1
Year   2017 2016 2015 2014

Illustrations by Pantek Media
Typeset by Pantek Media, Maidstone, Kent
Printed in Dubai

A catalogue record for this title is available from the British Library

ISBN 978 1 444 19379 4

Contents

Numbers

1 Numbers and Place Value

Digits and Place Value

The numbers that we use every day are made up of **digits**. The digits are:

0   1   2   3   4   5   6   7   8   9

The place of each digit in a number tells us the value of that digit. This is called the **place value** of the digit.

In the number 246 we can see how much each digit is worth from its place in the number. The digit on the right-hand side gives the number of ones (1s) or **units**. The digit next to it on the left gives the number of **tens** (10s), and the next digit to the left of that gives the number of **hundreds** (100s). We can use the letters "h", "t" and "u" to stand for hundreds, tens and units.

This table shows the place value of each digit in the number 246.

Hundreds	Tens	Units
h	t	u
2	4	6
This digit has the value of 2 hundreds:	This digit has the value of 4 tens:	This digit has the value of 6 units:
200	40	6

The number 246 can be split into:

200 *and* 40 *and* 6 = 246

Every time we move to the left in a number, the place value of the digit is ten times bigger than the place value of the digit before.

Exercise 1

For each of the numbers below write down the value of the underlined digit. The first one is done for you as an example:

<u>2</u>6   The answer is 20.

a) <u>3</u>2

b) 42<u>5</u>

c) <u>3</u>29

d) <u>5</u>63

For bigger numbers we can keep adding new columns to the left. These columns give the number of **thousands**, then **ten thousands**, then **hundred thousands**, and so on.

So the number 256 932 can be written as shown in this table.

Hundred thousands	Ten thousands	Thousands	Hundreds	Tens	Units
hth	tth	th	h	t	u
2	5	6	9	3	2
This digit has the value of 2 hundred thousands:	This digit has the value of 5 ten thousands:	This digit has the value of 6 thousands:	This digit has the value of 9 hundreds:	This digit has the value of 3 tens:	This digit has the value of 2 units:
200 000	50 000	6000	900	30	2

The number 256 932 can be split into:

200 000 *and* 50 000 *and* 6000 *and* 900 *and* 30 *and* 2 = 256 932

In large numbers we often use spaces between each group of three digits to make them easier to read. So a number like 999999 is often written as 999 999.

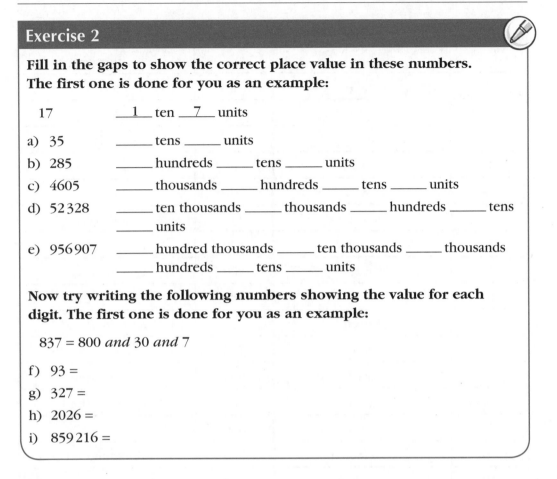

Exercise 2

Fill in the gaps to show the correct place value in these numbers. The first one is done for you as an example:

17   __1__ ten __7__ units

a) 35   _____ tens _____ units

b) 285   _____ hundreds _____ tens _____ units

c) 4605   _____ thousands _____ hundreds _____ tens _____ units

d) 52 328   _____ ten thousands _____ thousands _____ hundreds _____ tens _____ units

e) 956 907   _____ hundred thousands _____ ten thousands _____ thousands _____ hundreds _____ tens _____ units

Now try writing the following numbers showing the value for each digit. The first one is done for you as an example:

837 = 800 *and* 30 *and* 7

f) 93 =

g) 327 =

h) 2026 =

i) 859 216 =

Whole Numbers

All the numbers that are made up from one or more complete units are known as **whole numbers**. Later in the book we will look at numbers that are not whole numbers, called fractions and decimals.

It is a good idea to practise the way we say whole numbers and write them out in words.

We can use the table on page 4 to help us work out how we would say and write a number in words.

Number	Name	Number	Name
1	one	30	thirty
2	two	40	forty
3	three	50	fifty
4	four	60	sixty
5	five	70	seventy
6	six	80	eighty
7	seven	90	ninety
8	eight	100	one hundred
9	nine	200	two hundred
10	ten	300	three hundred
11	eleven	400	four hundred
12	twelve	500	five hundred
13	thirteen	600	six hundred
14	fourteen	700	seven hundred
15	fifteen	800	eight hundred
16	sixteen	900	nine hundred
17	seventeen	1000	one thousand
18	eighteen	5000	five thousand
19	nineteen	10 000	ten thousand
20	twenty	100 000	one hundred thousand

So the number 73 would be written in words as "seventy-three". Here are some more examples:

27 is written in words as "twenty-seven".
91 is written in words as "ninety-one".
236 is written in words as "two hundred and thirty-six".
1962 is written in words as "one thousand nine hundred and sixty-two".

Exercise 3

Write these numbers in words. You can look at the table on page 4 to help you.

a) 17   b) 23

c) 38   d) 92

e) 352   f) 2417

Ordering Numbers

Numbers often need to be placed in order of size from the smallest to the biggest or from the biggest to the smallest. We can tell by looking at the place values which number is biggest. 526 and 583 both have the same value for the hundreds (5), but 583 has a bigger value for the tens (583 has 8 tens, 526 has 2 tens). So 583 is bigger than 526. Other words for "bigger" are "larger" and "greater".

Exercise 4

Try answering the following questions to practise putting numbers in the right order.

a) Which number is bigger, 2439 or 2453?

b) Put these numbers in order from the smallest to the biggest.

527, 469, 1093, 562, 402, 1126

c) Put these numbers in order from the largest to the smallest.

127, 836, 453, 1100, 357, 1099

The Number Line

We can also think of numbers as points on a line. This line is called a **number line**. Number lines can be very useful for placing numbers in order of size. They can also help us with addition and subtraction.

Numbers on a number line always get bigger as we move to the right and smaller as we move to the left. So any number on a number line is always bigger than all the numbers to its left and always smaller than all the numbers to its right. Look at the number line on page 5. We can see that 9 is bigger, or "greater", than all the numbers to its left and smaller, or "less", than all the numbers to its right.

In mathematics we have special signs that mean "greater than" and "less than". The signs we use are > for "greater than" and < for "less than". For example:

"9 is greater than 7" can be written as $9 > 7$
"4 is less than 6" can be written as $4 < 6$

Exercise 5

Look at the number line on page 5. Are the following true or false?

a) $5 > 4$ b) $3 < 1$

c) $9 > 8$ d) $6 > 7$

e) $4 < 2$

Positive and Negative Numbers

Positive numbers are numbers that are bigger than 0. **Negative numbers** are numbers smaller than 0. We show negative numbers by putting a − sign in front of them. Negative numbers are placed on the number line to the left of 0.

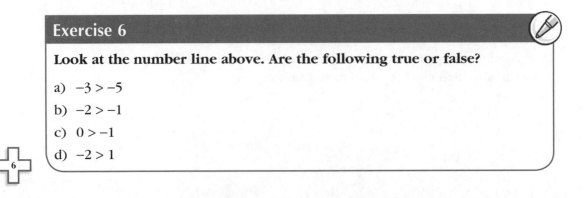

Remember that any number on a number line is always bigger than all the numbers to its left and smaller than all the numbers to its right. So we can see from the number line that −3 is bigger (or "greater") than −4, but smaller (or "less") than −2. We can write these facts as $-3 > -4$ and $-3 < -2$.

Exercise 6

Look at the number line above. Are the following true or false?

a) $-3 > -5$

b) $-2 > -1$

c) $0 > -1$

d) $-2 > 1$

Remember!

◆ The place of a digit in a number tells us the value of the digit.

◆ Every time we move one place to the left in a number the place value gets ten times bigger.

◆ Numbers on a number line always get bigger as we move to the right and smaller as we move to the left.

◆ Positive numbers are bigger than 0 and negative numbers are smaller than 0.

Revision Test on Numbers and Place Value

Now that you have worked your way through the chapter, try this revision test. The answers are in the answer book.

1. Write down the value of the underlined digit in 48<u>5</u>4.

2. Write down the value of the underlined digit in 32<u>7</u>.

3. Write down the value of the underlined digit in 8<u>3</u> 325.

4. Write down the value of the underlined digit in <u>7</u>63 211.

5. Write down the value of the underlined digit in 362 0<u>1</u>0.

6. Write this number in words: 93

7. Write this number in words: 510

8. Write this number in words: 2009

9. Write this number in words: 14 020

10. Write this number in digits: six hundred and fifty-four

11. Write this number in digits: seventeen thousand and twelve

12. Write this number in digits: twenty-six thousand eight hundred and thirty-two

13. Put these numbers in order of size from the smallest to the largest:
413, 359, 1053, 511, 403, 1105

14. Put these numbers in order of size from the largest to the smallest:
232 061, 241 052, 97 836, 17 901, 9882, 42 001

15. Put these numbers in order of size from the smallest to the largest:
−2, 0, 3, −5, 5, −1

16. Put these numbers in order of size from the largest to the smallest:
−7, 4, 1, 6, 2, −2

17. −8 > 7   True or false?

18. 5 < 4   True or false?

→

Revision Test on Numbers and Place Value *(continued)*

19. −17 > −18   True or false?

20. 0 > −1   True or false?

21. −3 < −2   True or false?

22. Sam has four cards. Each card has a number written on it. The four numbers are 2, 5, 0, 6. With these cards he can make this four-digit number: 2065. What is the largest four-digit number Sam can make from these four cards?

23. Using two of the same four cards (2, 5, 0, 6), what is the largest two-digit number Sam can make?

24. Make a number between 4000 and 4500 using these four digits: 5, 4, 7, 1

25. With the same four digits (5, 4, 7, 1) make a number between 7500 and 8000.

2 Addition and Subtraction

What is Addition?

Addition is another word for adding. We use the sign + to show that we want to add numbers together. We can use a number line to add numbers together by "counting on". If we want to work out 5 + 3 we can count on using the number line to help us like this:

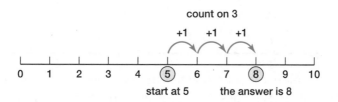

Additions give the same answer whichever order we do them. So 5 + 3 gives the same answer as 3 + 5. Because additions give the same answer whichever order we do them, we can start with the bigger number and add on the smaller number. This is easier than starting with the smaller number and adding on a bigger number. For bigger numbers we can change the number line to help us with the question. If we want to work out 14 + 63 we can use this number line to work out the answer:

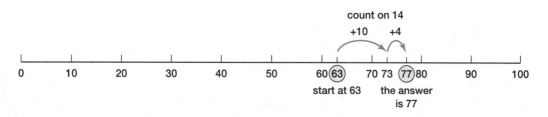

With practice, we soon don't need to draw a number line every time. We can add numbers together in our heads. As well as using a number line there are other things we can do to help us add numbers together in our heads. Here are some useful ideas:

1. Split the numbers up into smaller numbers. 23 + 35 is the same as 20 + 3 + 30 + 5. We can then think of this as 30 + 20 + 5 + 3 = 58.
2. Count on from the biggest number. 56 + 30 is the same as 56 + 10 + 10 + 10 = 86.
3. Use "nearly numbers" if the numbers are nearly the same as a number that is easier to work with. 21 + 12 is nearly 20 + 12. It is easier to add 20 + 12 so we can do this first and then add one more. 21 + 12 = 20 + 12 + 1 = 33.

When we write out a sum such as 21 + 12 = 33, we call it a "number sentence".

Exercise 1

Try the following addition questions in your head. If you get stuck, draw a number line to help you.

a) $2 + 3 =$

b) $5 + 6 =$

c) $10 + 7 =$

d) $21 + 9 =$

e) $6 + 5 + 2 =$

f) $35 + 31 =$

g) $45 + 52 =$

h) $21 + 73 =$

i) $51 + 53 + 58 =$

When numbers get too big to add together in our heads we can work out the answer by writing down the question and using a **written method**.

Here is $378 + 254$ worked out in three different ways using written methods. You should try all three methods to see which way works best for you.

First Written Method

Split the numbers into hundreds, tens and units and then add the columns.

$378 + 254$

$$
\begin{array}{r}
300 + 70 + 8 \\
200 + 50 + 4 \\
\hline
500 + 120 + 12 = 632
\end{array}
$$

Second Written Method

Use place value columns to add the units, then the tens, then the hundreds, then add them all together.

$378 + 254$

	h	t	u
	3	7	8
+	2	5	4
$8 + 4 =$		1	2
$70 + 50 =$	1	2	0
$300 + 200 =$	5	0	0
	6	3	2

Third Written Method

Put the digits into place value columns and then add them together vertically.

378 + 254

```
  h  t  u
  3  7  8
+ 2  5  4
  1  1
 ─────────
  6  3  2
```

First we add the units: $8 + 4 = 12$. We know that 12 is made up of 2 units and 1 ten, so we write 2 in the answer row under the units and we add the 1 ten to the tens column. We write a small 1 in the tens column to remind us to add this. Some people call this "carrying" the ten.

Then we add the tens: $7 + 5 = 12$, plus the 1 ten we have added from the units, $12 + 1 = 13$. 13 tens is made up of 3 tens plus 10 tens. 10 tens is 1 hundred so 13 tens is the same as 3 tens and 1 hundred. So we write 3 in the answer row under the tens and we add the 1 hundred to the hundreds column. We write a small 1 in the hundreds column to remind us to add this.

Then we add the hundreds: $3 + 2 = 5$, plus the 1 hundred we have added from the tens, $5 + 1 = 6$. So we write 6 in the answer row under the hundreds.

So the answer is 632.

Here is another addition worked out in the same way:

347 + 182

```
  h  t  u
  3  4  7
+ 1  8  2
  1
 ─────────
  5  2  9
```

First we add together the units: $7 + 2 = 9$. So we write 9 in the answer row under the units.

Then we add the tens: $4 + 8 = 12$. So we have 12 tens. 12 tens is made up of 2 tens plus 10 tens. 10 tens is 1 hundred so 12 tens is the same as 2 tens plus 1 hundred. So we write 2 in the answer row under the tens, and add the 1 hundred to the hundreds column. We write a small 1 in the hundreds column to remind us to add this.

Then we add the hundreds: $3 + 1 = 4$, plus 1 hundred we have added from the tens, $4 + 1 = 5$.

So the answer is 529.

Lots of people find carrying numbers into different columns difficult, but practice makes it easier.

Remember, there are lots of different ways of working out addition questions.

Exercise 2

Try these addition questions, using any method that works for you.

a) $72 + 112 =$

b) $128 + 153 =$

c) $372 + 247 =$

d) $4352 + 865 =$

e) $289 + 721 =$

What is Subtraction?

Subtraction is often called "taking away". We can use a number line for subtraction, but instead of counting on, we count back. If we want to work out $5 - 3$ we can count back using the number line like this:

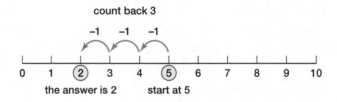

Subtractions don't give the same answer whichever way we do them. $5 - 3$ does not give the same answer as $3 - 5$. So we must always work out a subtraction in the order it is written.

There are things we can do to help us work out subtractions in our heads. Here are two ideas:

1. Count back. $45 - 13$ is the same as $45 - 10 - 3 = 32$.
2. Use "nearly numbers" if the numbers are nearly the same as a number that is easier to work with. $35 - 19$ is nearly $35 - 20$, which is easier to work out in our heads. We then need to add one back on because we subtracted one too many. So $35 - 19$ is the same as $35 - 20 + 1 = 16$.

Exercise 3

Try the following subtraction questions in your head. If you get stuck, draw a number line to help you.

a) $7 - 4 =$

b) $17 - 9 =$

c) $58 - 13 =$

d) $86 - 19 =$

e) $123 - 48 =$

Adding and Subtracting are Opposites

Adding and subtracting are opposites. You can use this fact to check your answer. For example, if you think that $17 - 9 = 8$, you can check it by adding 9 to 8 to see if you get back to 17.

$17 - 9 = 8$

Check with addition: $8 + 9 = 17$

Subtracting Bigger Numbers

When numbers get too big to subtract in our heads we can use place value to subtract in columns.

So $685 - 273$ can be set out like this:

```
  h t u
  6 8 5
− 2 7 3
  4 1 2
```

We start with the units and subtract 3 from 5, which leaves 2. We write this in the answer row under the units.

Then we move to the tens and subtract 7 from 8, which leaves 1. We write this in the answer row under the tens.

Finally we move to the hundreds and subtract 2 from 6, which leaves 4. We write this in the answer row under the hundreds.

So the answer is 412.

Sometimes the digit on the bottom line is bigger than the digit above it on the top line. When this happens we have to take 1 from the top row of the column to the left and add it to the top row of the column we are working in.

Look at this subtraction:

532 − 243

```
    h    t    u
   ⁴5̷  ¹²3̷  ¹2
 −  2    4    3
   ─────────────
    2    8    9
```

We start with the units: 3 is bigger than 2 so we take 1 ten from the top row of the tens column and we give it to the units column. Our 2 now becomes 12 because we have added 10 from the tens column. 12 − 3 = 9 so we write 9 in the answer row.

Now we deal with the tens: because we have given 1 ten from the tens column to the units column, we now have 2 tens left. So our subtraction is now 2 − 4. Because 4 is greater than 2, we must take 1 hundred from the top row of the hundreds column and give it to the top row of the tens column. 1 hundred is the same as 10 tens, so the 2 in the tens column now becomes 12 because we have added 10 tens to 2 tens. So now we have 12 − 4 = 8 in the tens column, and we write 8 in the answer row.

Finally we move to the hundreds: as 1 hundred has been given to the tens column, we are left with 4 − 2 = 2.

So the answer is 289.

Remember, we could check our answer by adding 289 to 243 to see if we get back to 532:

532 − 243 = 289

289 + 243 = 532

You can check this by using any of the methods we have learnt for addition.

Sometimes we find that there is a 0 in the top line. For example:

604 − 147

```
    h    t    u
   ⁵6̷  ⁹1̷0̷  ¹4
 −  1    4    7
   ─────────────
    4    5    7
```

First we look at the units: $4 - 7$ means we need to take a ten from the top row of the tens column. But there aren't any tens there. So first we need to move 1 hundred to the tens column to make it 10 tens. We now have 10 tens in the tens column. Next we give one of those tens to the units, leaving 9 tens in the top row of the tens column and 14 units in the top row of the units column. So now we have $14 - 7 = 7$ in the units column.

Next we move to the tens: we now have $9 - 4 = 5$.

Finally, in the hundreds column, we now have $5 - 1 = 4$.

So the answer is 457.

Difference

Subtraction helps us to work out the **difference** between two numbers. If Sally has 27 pencils and Darren has 13 pencils we can work out the difference in the number of pencils they have by subtracting Darren's 13 pencils from Sally's 27 pencils:

$27 - 13 = 14$

So Sally has 14 more pencils than Darren.

Exercise 4

Try these subtraction questions, using any method that works for you.

a) $853 - 621 =$ b) $3426 - 544 =$

c) $602 - 473 =$ d) $5321 - 322 =$

e) $758 - 86 =$

Remember!

- Adding gives the same answer whichever order we do it, but subtracting doesn't. We must always answer subtraction questions in the order they are written.
- We can use a number line to help us add and subtract.
- Adding and subtracting are opposites.
- Subtraction helps us to work out the difference between two numbers.
- There are lots of different ways to work out addition and subtraction questions. Always use the way that works best for you.

Revision Test on Addition and Subtraction

Now that you have worked your way through the chapter, try this revision test. The answers are in the answer book.

1. $39 + 14 =$

2. $39 - 14 =$

3. $53 + 27 + 19 =$

4. $226 + 341 =$

5. $344 + 237 =$

6. $765 - 532 =$

7. $826 - 437 =$

8. $1543 + 268 =$

9. $703 - 235 =$

10. $2457 + 156 =$

11. $1648 - 216 =$

12. Fill in the missing number: $17 + \underline{\hspace{1cm}} = 35$

13. Fill in the missing number: $1240 - \underline{\hspace{1cm}} = 540$

14. Fill in the missing number: $\underline{\hspace{1cm}} - 14 = 69$

15. Fill in the missing sign (+ or −): $73 - 12 \underline{\hspace{1cm}} 5 = 66$

16. Fill in the missing sign (+ or −): $1011 \underline{\hspace{1cm}} 11 + 153 = 1153$

17. Fill in the missing signs (+ or −): $92 \underline{\hspace{1cm}} 7 \underline{\hspace{1cm}} 19 = 80$

18. A shop has seven cans of meat, 19 cans of fruit, and 12 cans of vegetables. How many cans does the shop have altogether?

19. Petal is 11 years old. Her brother is 18 years old. How many years younger than her brother is Petal?

20. A town has a primary school and a secondary school. The primary school has 418 pupils, the secondary school has 215 pupils. How many more pupils are there in the primary school than in the secondary school?

21. Which two of these numbers have a difference of 5?
    3, 11, 15, 12, 4, 10

22. Which two of these numbers have a difference of 9?
    1, 4, 9, 13, 6, 19

23. Which two of these numbers have a difference of 4?
    −2, 0, 5, −1, 2, 11

24. James scored 26 goals in the season. Matthew scored 35 goals. What is the difference in the number of goals scored?

25. Jamal's mother is 32 years old. His grandmother is 58 years old. What is the difference in age between Jamal's mother and grandmother?

3 Multiplication

What is Multiplication?

Multiplication is the same as adding the same number together lots of times until we have added it together as many times as we need. For example, if we want to add together three 4s we do the following addition:

$$4 + 4 + 4 = 12$$

So three 4s added together equals 12. A shorter way of writing this is as a multiplication, like this:

$$4 \times 3 = 12$$

You will often hear a multiplication like this said out loud as "four times three equals 12" or "three fours equal 12". Another way to say it is "three lots of four equals 12".

Here is another example. This time we want to add together four 2s:

$$2 + 2 + 2 + 2 = 8$$

This is the same as writing:

$$2 \times 4 = 8$$

An important thing to remember about multiplications is that they give the same answer whichever order you write them. For example, "three lots of four equals 12" and also "four lots of three equals 12":

$4 \times 3 = 12$ this is the same as $4 + 4 + 4 = 12$
and
$3 \times 4 = 12$ this is the same as $3 + 3 + 3 + 3 = 12$

Exercise 1

Try these for yourself.

a) $2 \times 3 =$

b) $5 \times 2 =$

c) $4 \times 5 =$

d) $3 \times 6 =$

e) $2 \times 7 =$

Writing out multiplications as additions is fine for small numbers but it takes a lot of time for bigger numbers such as 9×8 or 8×7. To save time we have multiplication tables, like the one on page 18. You can save yourself a lot of time by learning your multiplication tables, but always remember what it is that you are learning. For example, on the multiplication table below you can see that $7 \times 4 = 28$. If you can't see this, then find 7 in the first column on the left-hand side and 4 in the top row, and then run one finger across from 7 and one finger down from 4. Where your fingers meet is the answer to 7×4.

You can learn and remember that $7 \times 4 = 28$, but don't forget that what it really means is:

$7 + 7 + 7 + 7 = 28$

The Multiplication Table

×	1	2	3	4	5	6	7	8	9	10	11	12
1	1	2	3	4	5	6	7	8	9	10	11	12
2	2	4	6	8	10	12	14	16	18	20	22	24
3	3	6	9	12	15	18	21	24	27	30	33	36
4	4	8	12	16	20	24	28	32	36	40	44	48
5	5	10	15	20	25	30	35	40	45	50	55	60
6	6	12	18	24	30	36	42	48	54	60	66	72
7	7	14	21	28	35	42	49	56	63	70	77	84
8	8	16	24	32	40	48	56	64	72	80	88	96
9	9	18	27	36	45	54	63	72	81	90	99	108
10	10	20	30	40	50	60	70	80	90	100	110	120
11	11	22	33	44	55	66	77	88	99	110	121	132
12	12	24	36	48	60	72	84	96	108	120	132	144

Exercise 2

Using the table above, look up the answers to the following multiplication questions:

a) $9 \times 6 =$ b) $3 \times 5 =$

c) $4 \times 9 =$ d) $8 \times 7 =$

e) $7 \times 8 =$

Remember that multiplications are the same whichever order you write them. So in Exercise 2 the answers to the last two questions should be the same because 8×7 gives the same answer as 7×8.

There are lots of ways to learn your multiplication tables (some people call them "times tables"). You can practise at home or with a friend. You can test members of your family. You can learn songs and rhymes to help you remember.

Exercise 3

If you have learnt your multiplication tables, cover over the table and try the following in your head:

a) $4 \times 4 =$

b) $6 \times 3 =$

c) $5 \times 9 =$

d) $7 \times 7 =$

e) $8 \times 4 =$

Word Problems

Multiplication helps us to answer word problems like this one:

> There are five children in a family. Each child eats three bananas. How many bananas are eaten altogether?

We could add together 3 bananas 5 times, once for each child:

$3 + 3 + 3 + 3 + 3 = 15$

But it is quicker to work out 3 bananas multiplied by 5 children:

$3 \times 5 = 15$

Here is another example:

> A school classroom has seven rows of four desks. How many desks are there in the classroom?

The answer is 7 rows × 4 desks = 28 desks. We could say 4 desks × 7 rows = 28 desks. We would get the same answer because multiplications give the same answer whichever order we do them.

Exercise 4

Try to work out these word problems. They are all multiplications.

a) A shop owner gave six boxes of orange juice bottles to a school. Each box had eight bottles in it. How many bottles were there altogether?

b) A netball team played seven matches and scored five goals in every game. How many goals did they score altogether?

c) A library lends nine books every day for five days. How many books do they lend altogether?

Multiplying a Number by 10, 100, 1000

When we multiply a whole number by 10, the digits all move one place to the left and we write a 0 on the end. For example, if we want to multiply 43 by 10, we work it out like this:

h	t	u		h	t	u
	4	3	× 10 =	4	3	0

So 43 × 10 = 430.

Here is another example:

18 × 10

h	t	u		h	t	u
	1	8	× 10 =	1	8	0

So 18 × 10 = 180.

Exercise 5

Try these questions yourself.

a) 7 × 10 =

b) 45 × 10 =

c) 98 × 10 =

d) 36 × 10 =

e) 73 × 10 =

f) 11 × 10 =

When we multiply a whole number by 100, the digits all move two places to the left and we write two zeros on the end. For example, if we want to multiply 43 by 100, we work it out like this:

th	h	t	u		th	h	t	u
		4	3	× 100 =		4	3	0	0

So 43 × 100 = 4300.

We can see now that:

43 × 10 = 430
43 × 100 = 4300

Exercise 6

Try these questions yourself.

a) 7 × 100 =

b) 45 × 100 =

c) 98 × 100 =

d) 36 × 100 =

e) 73 × 100 =

f) 11 × 100 =

When we multiply a whole number by 1000, the digits all move three places to the left and we write three zeros on the end. For example, if we want to multiply 43 by 1000, we work it out like this:

tth	th	h	t	u			tth	th	h	t	u
		4	3		× 1000 =			4	3	0	0	0

So 43 × 1000 = 43 000.

We can see now that:

43 × 10 = 430
43 × 100 = 4300
43 × 1000 = 43 000

Can you see the pattern?

Exercise 7

Now try these questions.

a) 7 × 1000 =   b) 45 × 1000 =

c) 98 × 1000 =   d) 36 × 1000 =

e) 73 × 1000 =   f) 11 × 1000 =

There are some more questions like these for you to try in the revision test at the end of the chapter.

Multiplying by Zero

It is useful to know that whenever we multiply a number by 0 the answer is always 0. For example, 4 × 0 = 0. This is because 4 × 0 is the same as "zero lots of four" which will, of course, still be 0. So any number multiplied by 0 gives an answer of 0.

Multiplying Bigger Numbers

Knowing our multiplication tables is fine for numbers such as 7 × 8 or 6 × 3, but what happens if we have a multiplication such as 153 × 4? Unless we know the multiplication table for 153, we need a way of working this out. The different ways we use to work out multiplications of bigger numbers are known as **written methods**. The written methods we use are called short multiplication and long multiplication.

Short Multiplication

We use **short multiplication** when the multiplying number is less than 10, such as 153×4. Split the number into units, tens and hundreds and multiply each one by 4:

$$
\begin{array}{rccc}
 & \mathbf{h} & \mathbf{t} & \mathbf{u} \\
 & 1 & 5 & 3 \\
\times & & & 4 \\
\hline
3 \times 4 = & & 1 & 2 \\
50 \times 4 = & 2 & 0 & 0 \\
100 \times 4 = & 4 & 0 & 0 \\
\hline
\text{now add all them all together} & 6 & 1 & 2 \\
\end{array}
$$

Another way of doing this short multiplication is to write it out like this:

$$
\begin{array}{rccc}
 & \mathbf{h} & \mathbf{t} & \mathbf{u} \\
 & 1 & 5 & 3 \\
\times & & & 4 \\
 & {}^2 & {}^1 & \\
\hline
 & 6 & 1 & 2 \\
\end{array}
$$

We start with the units: $3 \times 4 = 12$. We put 2 in the answer row and then add (or "carry") the ten units remaining to the tens column as 1 ten. We write a small 1 in the tens column to show this. Then we multiply the 5 in the tens column by 4, and add the 1 ten we have carried from the units: $5 \times 4 = 20$, then $20 + 1 = 21$.

We write 1 in the answer line in the tens column and then carry the 2 to the hundreds column. We write a small 2 in the hundreds column to show this. We multiply the 1 in the hundreds column by 4, and then add the 2 we carried from the tens column: $1 \times 4 = 4$, then $4 + 2 = 6$. So we write 6 in the answer line in the hundreds column. This gives us the answer 612.

Exercise 8

Try the following multiplication questions using the short multiplication method that you like best:

a) $82 \times 3 =$

b) $162 \times 7 =$

c) $224 \times 4 =$

d) Each class in a school has 38 children. There are six classes altogether. How many children are there in the school?

Long Multiplication

When the multiplying number is bigger than 10, we use a **long multiplication** method. The three most common long multiplication methods are called the **grid method**, the **expanded method** and the **vertical method**.

Let's work out this multiplication using all three ways:

241×12

Using the Grid Method

We break down both numbers into hundreds, tens and units and put them into a grid like the one below. Then we multiply each column by each row, and finally we add up all the answers.

×	200	40	1
10	2000	400	10
2	400	80	2

$2400 + 480 + 12 = 2892$

Using the Expanded Method

When we use the expanded method, we break down one of the numbers into smaller numbers to make the multiplication easier. In this question, we can break down the 12 into $10 + 2$. Then we multiply 241 by 10 and also by 2, and add the answers together.

$$
\begin{aligned}
241 \times 12 &= 241 \times (10 + 2) \\
&= (241 \times 10) + (241 \times 2) \\
&= 2410 + 482 \\
&= 2892
\end{aligned}
$$

Using the Vertical Method

Although it looks different, the vertical method is very similar to the expanded method. We multiply the first number by the units and then by the tens, and then add the answers together to get the final answer.

		th	h	t	u
			2	4	1
	×			1	2
$241 \times 2 =$			4	8	2
$241 \times 10 =$		2	4	1	0
add the answers		2	8	9	2

Exercise 9

Now try the following questions using one of the long multiplication methods. Remember to always show your working because in a real test you might get a mark for showing that you know what to do even if you get the wrong answer.

a) $125 \times 15 =$

b) $214 \times 25 =$

c) $527 \times 12 =$

d) A new housing section is being built in a city. There are 18 new roads and each road has 58 new houses. How many new houses are being built altogether?

Estimating Answers

Whichever method you use to work out a multiplication, it is always a good idea to first work out a rough estimate of what the answer is going to be. For example, if you are asked to work out 347×9, then you know that $347 \times 10 = 3470$, and you know that $300 \times 10 = 3000$. So if your answer is not between 3000 and 3470, you know you have gone wrong somewhere. We can work out a high estimate, a low estimate, and a "good" estimate for problems, which will help us know whether our answer is likely to be correct.

For example, if we are asked to work out 143×16, we can say that:

a high estimate would be $150 \times 20 = 3000$
a low estimate would be $140 \times 10 = 1400$
a "good" estimate would be $140 \times 15 = 2100$

So we know that the answer is going to be close to 2100. Now work out the answer to 143×16 using a long multiplication method. Is the answer closer to 2100 than to 3000 or 1400?

Remember!

◆ Multiplication is a quick way of working out repeated addition.
◆ Multiplications give the same answer whichever order you write them. So 4×5 gives the same answer as 5×4.
◆ When multiplying big numbers, use one of the written methods.
◆ If we multiply a number by 10, the digits move one place to the left. If we multiply by 100 the digits move two places to the left. If we multiply by 1000, the digits move three places to the left.

Remember! *(continued)*

- If we multiply any number by 0, the answer will be 0.
- It is always a good idea to work out an estimate of the answer.
- There are lots of different ways of working out multiplication questions. Use the way that works best for you.

Revision Test on Multiplication

Now that you have worked your way through the chapter, try this revision test. The answers are in the answer book. Answer the questions without using a calculator.

1. $3 \times 4 =$
2. $7 \times 7 =$
3. $8 \times 6 =$
4. $32 \times 10 =$
5. $53 \times 10 =$
6. $18 \times 100 =$
7. $32 \times 100 =$
8. $21 \times 1000 =$
9. $7 \times 12 =$
10. $64 \times 4 =$
11. $43 \times 6 =$
12. $134 \times 3 =$
13. $146 \times 12 =$
14. $235 \times 11 =$
15. $87 \times 17 =$
16. $214 \times 25 =$
17. $691 \times 24 =$
18. James ate three packets of sweets in one day. Each packet had 16 sweets in it. How many sweets did he eat altogether? (Remember, the multiplication to work out is three packets of 16 sweets, which is 16×3.)
19. A school bought seven packs of pencils. Each pack had 25 pencils in it. How many pencils did the school buy altogether?
20. A shop sold 12 packs of biscuits in one day. There were 12 biscuits in each pack. How many biscuits were sold altogether? $\rightarrow$

Revision Test on Multiplication *(continued)*

21. Tennis balls are sold in boxes. Each box has six tennis balls in it. How many tennis balls are there in 35 boxes?

22. On a farm there are 85 goats. How many legs do the goats have altogether?

23. Mary buys some chalk for her school. In each box there are 40 sticks of chalk. Mary buys 21 boxes. How many sticks of chalk does she buy?

24. A train has 14 carriages. Each carriage can hold 92 people. How many people can travel on the train?

25. 3217 people live in a town. If seven times as many people lived in the town, how many people would live in the town altogether?

4 Division

What is Division?

Division is the opposite of multiplication. You will remember from Chapter 3 *Multiplication* that $4 \times 3 = 12$ is a short way of saying $4 + 4 + 4 = 12$. In other words, multiplication is a short way of working out a repeated addition. Division is a short way of working out a repeated subtraction.

For example, if we have 12 sweets and we want to share them equally between 4 pupils, we can keep giving each pupil a sweet until there are no sweets left. To start, each pupil gets 1 sweet, so we have subtracted 4 sweets from our total of 12:

$12 - 4 = 8$

There are still 8 sweets to share. Next, we give each pupil another sweet, so we subtract another 4 sweets from our total:

$12 - 4 - 4 = 4$

There are now 4 sweets left. Next, we give each pupil another sweet, so we subtract another 4 sweets from the total:

$12 - 4 - 4 - 4 = 0$

There are now no sweets left. So by giving each pupil 1 sweet 3 times we have used up all the sweets.

Each pupil has 3 sweets. So 12 sweets shared equally between 4 pupils = 3 sweets each. We write this as a division like this:

$12 \div 4 = 3$

The sign we often use for division is ÷, but $12 \div 4$ can also be written as:

$12/4$ or $\dfrac{12}{4}$ or $4\overline{)12}$

We will see some examples of divisions written like this later in the book.

You will often hear a division like $12 \div 4$ said out loud as "12 divided by four equals three" or "12 shared equally between four equals three".

Here is another example of a word problem that we can answer using division:

> We have 12 sweets. To how many people can we give two sweets each?

We can keep subtracting 2 sweets until there are no sweets left. We would do this 6 times, so the answer is that we can give 2 sweets to 6 people before we use up all the sweets:

$12 \div 2 = 6$

Another way to answer this question is to think about division as the opposite of multiplication, so we could work out how many people multiplied by 2 sweets each equals 12 sweets:

2 sweets × number of people = 12 sweets

You know from your multiplication tables that $2 \times 6 = 12$, so the answer to our problem is 6 people.

You will often hear people say a division like $12 \div 2 = 6$ out loud as "two goes into 12 six times".

Notice that $2 \times 6 = 12$ and $12 \div 2 = 6$ have exactly the same numbers in them. This is because multiplication and division are opposites. One important difference is that multiplications give the same answer whichever order we write them.

$4 \times 2 = 8$ and $2 \times 4 = 8$

This is not true for division. $4 \div 2$ does not give the same answer as $2 \div 4$. So we must always work out a division in the order it is written.

We now know that additions and multiplications give the same answer whichever order they are written. Subtractions and divisions do not give the same answer whichever order they are written so we must always work them out in the order given in the question.

Because multiplication and division are opposites, we can think of any multiplication or division and write three different multiplication and division number sentences just using those numbers. For example, if we choose $2 \times 5 = 10$ we can also write:

$5 \times 2 = 10$, $10 \div 2 = 5$ and $10 \div 5 = 2$

Here is another one. We know that 3, 5 and 15 are linked because we know that $3 \times 5 = 15$. This means that the following number sentences are also true:

$5 \times 3 = 15$
$15 \div 3 = 5$
$15 \div 5 = 3$

Exercise 1

Write four correct number sentences using multiplication and division for each of these sets of numbers.

a) 3, 4, 12   b) 3, 9, 27

c) 5, 7, 35   d) 7, 8, 56

e) 2, 8, 16   f) 5, 9, 45

Because division is the opposite of multiplication, we can use a multiplication table to work out the answers to division problems.

×	1	2	3	4	5	6	7	8	9	10	11	12
1	1	2	3	4	5	6	7	8	9	10	11	12
2	2	4	6	8	10	12	14	16	18	20	22	24
3	3	6	9	12	15	18	21	24	27	30	33	36
4	4	8	12	16	20	24	28	32	36	40	44	48
5	5	10	15	20	25	30	35	40	45	50	55	60
6	6	12	18	24	30	36	42	48	54	60	66	72
7	7	14	21	28	35	42	49	56	63	70	77	84
8	8	16	24	32	40	48	56	64	72	80	88	96
9	9	18	27	36	45	54	63	72	81	90	99	108
10	10	20	30	40	50	60	70	80	90	100	110	120
11	11	22	33	44	55	66	77	88	99	110	121	132
12	12	24	36	48	60	72	84	96	108	120	132	144

Find the number 40 in the 8-column. If you run your finger down the column from the 8 at the top, you will find it. If you look back along the row, you will see that this is the 5-row. We know from multiplication that this means $5 \times 8 = 40$ and $8 \times 5 = 40$. But it also means that $40 \div 8 = 5$ and $40 \div 5 = 8$. So if we are asked the question "What is $40 \div 5$?" we could use the multiplication table to find out.

Let's work out $108 \div 12$ using the multiplication table. First we look down the 12-column until we find 108. Then we look back along the row to see which row we are in. The answer is 9. So $108 \div 12 = 9$.

Exercise 2

Now try the following division problems. If you get stuck, use the multiplication table to help you.

a) $25 \div 5 =$

b) $18 \div 3 =$

c) $56 \div 7 =$

d) $60 \div 10 =$

e) $96 \div 8 =$

f) $36 \div 6 =$

g) Lucy has 16 cakes. She wants to share them equally between her four brothers. How many cakes does she give each brother?

There are some more questions like these for you to try in the revision test at the end of the chapter.

Remainders

Sometimes division problems don't have a whole number as an answer.
Think about this problem:

$9 \div 2$

We can work out that $8 \div 2 = 4$. So how many 2s are there in 9? The answer is 4 with 1 left over, or remaining. We write down the answer to this problem like this:

$9 \div 2 = 4 \text{ r}1$

The answer is "4 remainder 1". The r stands for "remainder".

Here is another example:

$16 \div 7$

We can work out that $14 \div 7 = 2$ but now there are two left over to get us to 16, so the answer is 2 remainder 2.

$16 \div 7 = 2 \text{ r}2$

Exercise 3

Try the following division questions. All the answers have remainders.

a) $7 \div 3 =$ b) $23 \div 7 =$

c) $11 \div 4 =$ d) $19 \div 5 =$

e) $15 \div 6 =$ f) $53 \div 7 =$

In Chapter 7 *Decimals*, we will learn how we can show the remainder in a different way.

We often have to work with remainders when answering problems. Here is an example of a word problem with a remainder:

John needs to put 18 cartons of orange juice into boxes. Each box holds four cartons. How many boxes does John need?

To find the answer, we divide 18 by 4 to see how many boxes of 4 cartons John needs to carry all the 18 cartons. This gives us:

$18 \div 4 = 4 \text{ r}2$

So John will fill 4 boxes with 4 cartons each and have 2 cartons left over (the remainder). John will need one more box to carry these 2 cartons, even though it won't be full. So the answer to the problem is 5 boxes.

Odd and Even

If we can divide a number by 2 and get an answer with no remainders, then it is an **even number**. If we can't divide a number by 2, then it is an **odd number**. For example, 2, 4, 6, 8 and 10 are all even numbers. So are 24, 26 and 28. But 3, 5, 7, 9 and 11 are all odd numbers, and so are 37, 73 and 101.

Exercise 4

Are the following numbers odd or even?

a) 17

b) 36

c) 13

d) 41

e) 18

f) 88

Dividing by 10, 100, 1000

You will remember from Chapter 3 *Multiplication* that when we multiply a whole number by 10 all the digits move one place to the left and we write a 0 on the end. Division is the opposite of multiplication, so when we divide a whole number by 10 all the digits move one place to the right.

For example, if we want to work out $230 \div 10$ we do it like this:

h	t	u		h	t	u
2	3	0	$\div 10 =$		2	3

So $230 \div 10 = 23$.

Here is another example:

$970 \div 10$

h	t	u		h	t	u
9	7	0	$\div 10 =$		9	7

So $970 \div 10 = 97$.

Exercise 5

Here are some questions for you to try.

a) $180 \div 10 =$

b) $820 \div 10 =$

c) $410 \div 10 =$

d) $100 \div 10 =$

e) $550 \div 10 =$

f) $1420 \div 10 =$

When we divide by 100, the digits all move two places to the right. For example:

2300 ÷ 100

th	h	t	u			th	h	t	u
2	3	0	0	÷ 100 =				2	3

So 2300 ÷ 100 = 23.

Exercise 6

Try these questions to practise dividing by 100.

a) 1800 ÷ 100 =   b) 8200 ÷ 100 =

c) 900 ÷ 100 =   d) 6000 ÷ 100 =

e) 5500 ÷ 100 =   f) 14 200 ÷ 100 =

When we divide by 1000, the digits all move three places to the right. For example:

23 000 ÷ 1000

tth	th	h	t	u			tth	th	h	t	u
	2	3	0	0	0	÷ 1000 =				2	3

So 23 000 ÷ 1000 = 23.

We can now see that:

230 ÷ 10 = 23
2300 ÷ 100 = 23
23 000 ÷ 1000 = 23

Can you see the pattern?

Exercise 7

Try these questions to practise dividing by 1000.

a) 18 000 ÷ 1000 =   b) 60 000 ÷ 1000 =

c) 1000 ÷ 1000 =   d) 146 000 ÷ 1000 =

e) 55 000 ÷ 1000 =   f) 85 000 ÷ 1000 =

Making Division Problems Easier

An important thing to know about division problems is that if we multiply both numbers by the same number we will get the same answer to the division. For example, if we take $8 \div 2$ we can see that the divisions below all give the same answer because in each case we have multiplied both the 8 and the 2 by the same number.

$8 \div 2$	$= 4$	
$16 \div 4$	$= 4$	we have multiplied both 8 and 2 by 2
$40 \div 10$	$= 4$	we have multiplied both 8 and 2 by 5
$80 \div 20$	$= 4$	we have multiplied both 8 and 2 by 10
$800 \div 200$	$= 4$	we have multiplied both 8 and 2 by 100

Knowing that we can do this to work out the answer to division problems can make working out some problems much easier.

For example, if we need to work out $75 \div 5$ we could multiply both numbers by 2 so that the problem becomes $150 \div 10$. This is much easier to work out. The answer to $150 \div 10$ is 15, so we know that the answer to $75 \div 5$ is also 15.

In the same way, if we divide both the numbers in a division by the same number it will give the same answer. This can also help us to solve some problems. For example, if we want to work out $96 \div 8$ we could divide both numbers by 2 to give $48 \div 4$. This is a bit easier but we can make it even easier by dividing both numbers by 2 again so that the problem becomes $24 \div 2$. This is much easier to work out in our heads.

$96 \div 8$ is the same as $48 \div 4$, which is the same as $24 \div 2 = 12$.

Exercise 8

Try these practice questions using any method that works for you.

a) $52 \div 4 =$ b) $120 \div 20 =$

c) $450 \div 5 =$ d) $115 \div 5 =$

e) $48 \div 8 =$ f) $300 \div 30 =$

Dividing Bigger Numbers

Sometimes division problems are too difficult to work out using the ideas we have used so far in this chapter. If we want to divide using bigger numbers then we can use **written methods** to help us.

The three most common written methods for division are repeated subtraction, short division and long division.

Repeated Subtraction

Repeated subtraction is a useful way to work out divisions. For example, if we want to work out $204 \div 6$ we could do it by taking away multiples of 6 until we get to the answer:

$204 \div 6$

$$
\begin{array}{r}
2\ 0\ 4 \\
-\quad 6\ 0 \\
\hline
1\ 4\ 4 \\
-\quad 6\ 0 \\
\hline
8\ 4 \\
-\quad 6\ 0 \\
\hline
2\ 4 \\
-\quad 2\ 4 \\
\hline
0
\end{array}
$$

10×6 (we have subtracted 10 sixes)

10×6 (we have subtracted another 10 sixes, making 20 sixes subtracted so far)

10×6 (we have subtracted another 10 sixes, making 30 sixes subtracted so far)

4×6 (we have subtracted another 4 sixes, making 34 sixes subtracted altogether)

We can see that we have subtracted 6 from 204 a total of 34 times altogether, and so $204 \div 6 = 34$.

Here is another example:

$350 \div 14$

$$
\begin{array}{r}
3\ 5\ 0 \\
-\quad 1\ 4\ 0 \\
\hline
2\ 1\ 0 \\
-\quad 1\ 4\ 0 \\
\hline
7\ 0 \\
-\quad 7\ 0 \\
\hline
0
\end{array}
$$

10×14

10×14

5×14

We can see that we have subtracted 14 from 350 a total of 25 times altogether and so $350 \div 14 = 25$.

Here is an example with a remainder:

$243 \div 12$

$$
\begin{array}{r}
2\ 4\ 3 \\
-\quad 1\ 2\ 0 \\
\hline
1\ 2\ 3 \\
-\quad 1\ 2\ 0 \\
\hline
3
\end{array}
$$

10×12

10×12

We can see that we have subtracted 12 from 243 a total of 20 times, leaving us with 3. We can't subtract any more 12s without going into negative numbers, so the answer is 20 remainder 3:

$243 \div 12 = 20 \text{ r}3$

Exercise 9

Try these problems using repeated subtraction.

a) $144 \div 4 =$

b) $280 \div 8 =$

c) $162 \div 6 =$

d) $322 \div 14 =$

e) $273 \div 7 =$

f) $597 \div 18 =$

g) A school has 108 new desks to share equally between nine classrooms. How many new desks should go into each classroom? (This question is $108 \div 9$, so we need to repeatedly subtract 9 from 108 to get the answer.)

Short Division

Short division is often used for dividing by a number less than 10. Some people like it better than repeated subtraction because it can be faster. For short division we use this sign $)$ to show that it is a division problem. So $84 \div 7$ would be written as $7\overline{)84}$.

To work out $7\overline{)84}$ using short division, we start at the left of the number and divide the 7 into the tens.

The number is 84 so there are 8 tens. 7 divides into 8 once, with 1 ten remaining. We write 1 above the line over the tens column to show that 7 divided into 8 once, and we add the one ten left over to the units column. This now makes 14 units altogether. We show this like this:

$$\frac{1}{7\overline{)8\ _14}}$$

Now we move to the units. We divide 14 by 7, which gives us 2. We show this like this:

$$\frac{1\ \ 2}{7\overline{)8\ _14}}$$

So $84 \div 7 = 12$.

Here is another example using short division:

$$6\overline{)6\ 7\ 8}$$

We start with the hundreds: there are 6 hundreds. 6 divides into 6 exactly once with nothing left over, so we write 1 above the line over the hundreds:

$$\frac{1}{6\overline{)6\ 7\ 8}}$$

Now we divide 6 into the number of tens: there are 7 tens, and 6 divides into 7 once with 1 remaining. We write 1 above the line over the tens to show that 6 divided into 7 once, and add the one ten left over to the units column. This now makes the units equal to 18:

$$\begin{array}{r} 1\ \ 1\ \ \\ \hline 6\overline{)6\ \ 7\,18} \end{array}$$

Then we look at the units: we divide 6 into 18, which is equal to 3:

$$\begin{array}{r} 1\ \ 1\ \ 3 \\ \hline 6\overline{)6\ \ 7\,18} \end{array}$$

So $678 \div 6 = 113$.

If the answer has a remainder, we show it in the usual way. For example, if the last problem was $679 \div 6$, the answer would be 113 remainder 1, which would look like this:

$$\begin{array}{r} 1\ \ 1\ \ 3\ \ \text{r1} \\ \hline 6\overline{)6\ \ 7\,19} \end{array}$$

Here is one more example, before you try some questions:

$$8\overline{)1\ \ 6\ \ 4\ \ 8}$$

Here, we must start with the thousands: 8 won't divide into 1, so we add the 1 thousand to the hundreds, so that there are now 16 hundreds altogether.

Now the hundreds: there are 16 hundreds. 8 divides into 16 exactly twice, with nothing left over to add to the tens column:

$$\begin{array}{r} 2\ \ \ \ \ \ \ \\ \hline 8\overline{)1\ \ \imath6\ \ 4\ \ 8} \end{array}$$

There are 4 tens. 8 won't divide into 4, so we write 0 above the line over the tens and add the 4 tens to the units:

$$\begin{array}{r} 2\ \ 0\ \ \ \ \\ \hline 8\overline{)1\ \ 6\ \ 4\ \ 48} \end{array}$$

Then we move to the units: there are now 48 units altogether. 8 divides into 48 exactly 6 times, so we write 6 above the line over the units:

$$\begin{array}{r} 2\ \ 0\ \ 6\ \\ \hline 8\overline{)1\ \ 6\ \ 4\ \ 48} \end{array}$$

So $1648 \div 8 = 206$.

Exercise 10

Answer these questions using short division.

a) $336 \div 3 =$ b) $228 \div 3 =$

c) $456 \div 4 =$ d) $238 \div 7 =$

e) $744 \div 6 =$ f) $898 \div 8 =$

g) 225 pencils need to be packed equally into five large boxes. How many pencils go into each box?

Long Division

Long division is sometimes used when we divide by a number greater than 10. It is like short division but is used for bigger numbers. Instead of adding any remaining thousands, hundreds or tens to the next column and showing them in the number itself, we write them below the problem and then bring the other digits down to them. It works like this.

Let's work out $2576 \div 23$ using long division:

$$23\overline{)2\ 5\ 7\ 6}$$

First we divide 23 into the thousands. There are 2 thousands so 23 won't divide into this. So now we divide 23 into the hundreds.

As we haven't used the thousands yet we have to include them in the hundreds, so there are now 25 hundreds. 23 divides into 25 once with 2 remaining. We show this by writing 1 above the 25 and 23 below it to show that 23 divided into 25 just once. Then we subtract the 23 from 25.

```
        1
23)2 5 7 6
    2 3
      2   we have subtracted 23 from 25 to find the remainder, which is 2
```

Next we bring down the tens digit like this:

```
        1
23)2 5 7 6
    2 3
      2 7
```

Now we divide 23 into 27. It goes once so we write 1 above the line over the tens and subtract one 23 from 27 to get the remainder:

```
        1 1
23)2 5 7 6
    2 3
      2 7
      2 3
        4     we have subtracted 23 from 27 to find the remainder, which is 4
```

Next we bring down the units digit like this:

```
        1 1
23)2 5 7 6
    2 3
      2 7
      2 3
        4 6
```

Now we divide 23 into 46. It goes in exactly twice. So we put 2 above the line over the units and we write 46 at the bottom because 23 divided into 46 *twice* and $23 \times 2 = 46$:

```
        1 1 2
23)2 5 7 6
    2 3
      2 7
      2 3
        4 6
        4 6     this is 23 × 2
        0 0     we can see that there is no remainder because 46 – 46 = 0
```

So the answer is 112:

$$2576 \div 23 = 112$$

It is worth remembering that we could have solved this problem by repeated subtraction like this:

```
      2 5 7 6
  −   2 3 0 0      100 × 23
        2 7 6
  −     2 3 0       10 × 23
          4 6
  −       4 6        2 × 23
            0
```

We can see that we have subtracted 23 from 2576 a total of 112 times, so the answer is 112:

$2576 \div 23 = 112$

Exercise 11

Work out the answers to these questions using long division.

a) $736 \div 23 =$ b) $3645 \div 15 =$

c) $1380 \div 12 =$ d) $7015 \div 23 =$

e) $3564 \div 27 =$ f) $437 \div 18 =$

g) 12 children share a bag of 216 sweets equally between them. How many sweets do they each get?

Remember!

◆ Division is the opposite of multiplication.

◆ Division is a quick way of working out repeated subtraction.

◆ Additions and multiplications give the same answer whichever order they are written. Subtractions and divisions do *not* give the same answer whichever order they are written so we must always work them out in the order given in the question.

◆ If we divide a number by 10, the digits move one place to the right. If we divide by 100, the digits move two places to the right. If we divide by 1000, the digits move three places to the right.

◆ There are lots of different ways to work out division problems.

Revision Test   on Division

Now that you have worked your way through the chapter, try this revision test. The answers are in the answer book. Answer the questions without using a calculator.

For each question use the division method that you think is best. You can choose between working out the problem in your head, repeated subtraction, short division or long division.

1. Write four correct number sentences, using multiplication and division, for each of these sets of numbers:

   a) 4, 16, 64

   b) 63, 9, 7

   c) 120, 3, 40

Revision Test　on Division *(continued)*

2.   a) $30 \div 6 =$

    b) $21 \div 3 =$

    c) $48 \div 8 =$

3.   a) $730 \div 10 =$

    b) $900 \div 100 =$

    c) $85\,600 \div 100 =$

    d) $6000 \div 100 =$

    e) $93\,425\,000 \div 1000 =$

4.   a) $360 \div 60 =$

    b) $125 \div 5 =$

    c) $959 \div 7 =$

    d) $2064 \div 3 =$

5.   Work out the following. Give the remainder in your answer.

    a) $319 \div 6 =$

    b) $59 \div 3 =$

    c) $147 \div 4 =$

    d) $1046 \div 9 =$

6.   A bag of 125 sweets is shared equally between five children. How many sweets do they get each?

7.   A football club has 21 footballs. The club wants to store the balls in bags. Each bag holds up to three balls. How many bags do they need?

8.   A teacher has a class with 36 pupils. She wants to give out textbooks so that each textbook is shared by three pupils. How many textbooks does she need?

9.   A farmer has 56 cows. He wants to transport them all to market in trucks. Each truck holds 20 cows. How many trucks will he need?

10. At a pencil factory, 1128 pencils are being packed into boxes. Each box will hold 24 pencils. How many boxes are needed?

5 More About Numbers

Multiples and Factors

Now that we know how multiplication and division work, it is useful to learn about multiples and factors.

Multiples

The **multiples** of a number are all the numbers we get when we multiply the number by 1, 2, 3, 4, 5 ... and so on:

The multiples of 2 are 2, 4, 6, 8, 10, 12 ... and so on.
The multiples of 3 are 3, 6, 9, 12, 15, 18 ... and so on.
The first five multiples of 7 are the first five numbers in the multiplication table for 7: 7, 14, 21, 28, 35

The multiples of 2 are the **even numbers**. All the other numbers that are not multiples of 2 are the **odd numbers**.

Lots of multiples have patterns that make them easier for us to see:

◆ **Multiples of 2 always end in 2, 4, 6, 8 or 0.**
   For example, we can see that 986 must be a multiple of 2 because it ends in 6. We can also see that 434 must be a multiple of 2 because it ends in 4.
◆ **Multiples of 5 always end in 5 or 0.**
   We can see that 45, 90 and 1455 must all be multiples of 5 because they all end in 5 or 0.
◆ **Multiples of 10 always end in 0.**
   For example, 30, 70, 380 and 4560 are all multiples of 10 because they all end in 0.
◆ **If the digits in a number add up to a multiple of 3 then the number is a multiple of 3.**
   For example, the digits in 162 add up to 9 (1 + 6 + 2 = 9). 9 is a multiple of 3, so we know that 162 must also be a multiple of 3.
◆ **If an even number is a multiple of 3 it must also be a multiple of 6.**
   162 is an even number and it is a multiple of 3, so it must also be a multiple of 6.

Exercise 1

a) Write down the first five multiples of 8.
b) Which of these numbers are multiples of 2?
   4, 7, 12, 23, 60, 87, 100, 113, 116
c) Which of these numbers are multiples of 5?
   15, 22, 45, 70, 153, 260, 335, 558

Exercise 1 *(continued)*

d) Which of these numbers are multiples of 10?

   15, 30, 65, 90, 122, 240, 755, 1000, 1020

e) Which of these numbers are multiples of 3?

   18, 33, 47, 54, 82, 123, 754, 1602

f) Which of the multiples of 3 in question e) are also multiples of 6?

Lowest Common Multiples

The multiples of 3 are 3, 6, 9, 12, 15, 18, 21, 24 …
The multiples of 4 are 4, 8, 12, 16, 20, 24, 28, 32 …

If we look at the multiples of 3 and 4 we can see that 12 and 24 appear in both lists. They are **common multiples** of 3 and 4.

12 is the smallest common multiple, so we call this number the **lowest common multiple** of 3 and 4. Lowest common multiples are very useful in mathematics.

Let's look at another example:

The multiples of 3 are 3, 6, 9, 12, 15, 18, 21, 24, 27, 30, 33 …
The multiples of 5 are 5, 10, 15, 20, 25, 30, 35, 40, 45, 50, 55 …

We can see that 15 and 30 are common multiples of 3 and 5 as they are in both lists, but 15 is the lowest common multiple.

Exercise 2

What is the lowest common multiple of each pair of numbers?

a) 2 and 3   b) 4 and 6

c) 5 and 10   d) 5 and 7

e) 6 and 15

Factors

The **factors** of a number are all the numbers that divide into it exactly. If we want to find the factors of 6, we must first think about which numbers divide into 6.

$6 \div 1 = 6$ and $6 \div 6 = 1$. So we can see that 1 divides exactly into 6 six times, and 6 divides exactly into 6 once. So 1 and 6 are factors of 6.

$6 \div 2 = 3$ and $6 \div 3 = 2$. So 2 and 3 must also be factors of 6.

No other numbers divide exactly into 6, so the factors of 6 are 1, 2, 3 and 6.

Because division is the opposite of multiplication, we can find factors by thinking about which numbers multiply together to make the number we are working with. For example, if we want to find the factors of 9, we know that $1 \times 9 = 9$, so 1 and 9 are factors of 9. We also know that $3 \times 3 = 9$, so 3 must also be a factor of 9. Are there any others? No, so the factors of 9 are 1, 3 and 9.

Here are some more examples of factors:

◆ The factors of 2 are 1 and 2.
◆ The factors of 3 are 1 and 3.
◆ The factors of 4 are 1, 2 and 4.
◆ The factors of 8 are 1, 2, 4 and 8.
◆ The factors of 20 are 1, 2, 4, 5, 10 and 20.

Exercise 3

Find all the factors of these numbers.

a) 5   b) 7

c) 10   d) 12

e) 15   f) 18

g) 25   h) 30

Highest Common Factors

It is often useful in mathematics to find the **highest common factor** of two numbers. For example, if we look at the factors of 8 and 20 we see that:

the factors of 8 are 1, 2, 4 and 8
the factors of 20 are 1, 2, 4, 5, 10 and 20.

We can see that both numbers have 1, 2 and 4 as factors. The biggest of these factors is 4, so the highest common factor of 8 and 20 is 4.

If we want to find the highest common factor of 6 and 15 we first write out the factors for each number:

the factors of 6 are 1, 2, 3 and 6
the factors of 15 are 1, 2, 3, 5 and 15

We can see that the highest common factor is 3.

Exercise 4

What is the highest common factor of each pair of numbers?

a) 6 and 9

b) 12 and 20

c) 15 and 30

d) 18 and 30

e) 9 and 27

f) 16 and 24

Prime Numbers

Prime numbers are special numbers that have only two factors. The two factors of a prime number are the number itself and 1. For example, 7 is a prime number as it can be divided only by 7 and 1. Eleven is a prime number, as it can be divided only by 11 and 1.

Here are all the prime numbers that are less than 20:

2, 3, 5, 7, 11, 13, 17, 19

The number 1 is not a prime number as it has only one factor, and all prime numbers have *two* factors.

The only even prime number is 2.

Exercise 5

Which of these numbers are prime numbers?

5, 9, 14, 19, 23, 28, 31, 35, 40, 41, 49

Why Are Prime Numbers Important?

Prime numbers are very important in computer programming and for writing secret codes! They are also important in nature. Some insects have life cycles that depend on prime numbers.

Square Numbers

If we multiply a whole number by itself we get a **square number**:

$1 \times 1 = 1$
$2 \times 2 = 4$
$3 \times 3 = 9$
$4 \times 4 = 16$

1, 4, 9 and 16 are all square numbers.

They are called square numbers because we can show how they are made by drawing squares, like these:

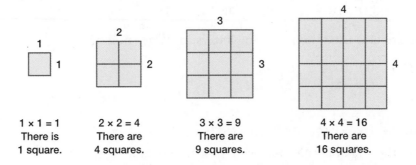

$1 \times 1 = 1$	$2 \times 2 = 4$	$3 \times 3 = 9$	$4 \times 4 = 16$
There is	There are	There are	There are
1 square.	4 squares.	9 squares.	16 squares.

We can write 2×2 like this: 2^2. We say this as "two squared".

3×3 is 3^2 and we say this as "three squared", and so on.

So:

$$1 \times 1 = 1^2 = 1$$
$$2 \times 2 = 2^2 = 4$$
$$3 \times 3 = 3^2 = 9$$
$$4 \times 4 = 4^2 = 16$$

Exercise 6

What are these squares equal to? The first one is done for you as an example:

$$2^2 = 4$$

a) 5^2 b) 6^2

c) 7^2 d) 8^2

e) 9^2 f) 10^2

g) 11^2 h) 12^2

i) $2^2 + 5^2$ j) $3^2 + 3^2$

Roman Numerals

What are Roman Numerals?

The numbers we use today were developed by people in the Middle East and India. Another way of writing numbers that is still sometimes used today was invented by the Romans. The Romans were from Rome in Italy. About 2000 years ago they ruled large parts of Europe and the Mediterranean. The Romans used letters for numbers. These letters are known as Roman numerals. Roman numerals are still used today on some clocks and watches, on buildings, and in books.

Roman numerals from 1 to 10 are written like this:

1	2	3	4	5	6	7	8	9	10
I	II	III	IV	V	VI	VII	VIII	IX	X

The Romans also used the following letters for larger numbers:

50	100	500	1000
L	C	D	M

How do we use Roman Numerals?

There are three simple rules for using Roman numerals:

1. We can't use more than three of the same letter together.
2. If there is a smaller number *before* a bigger number, we subtract the smaller number from the bigger number.
3. If there is a smaller number *after* a bigger number, we add the smaller number to the bigger number.

We can see how these rules go together by looking at the number 4. The first rule means that the Romans didn't write 4 as IIII. Instead they wrote it as IV. This means "V take away I" (or 5 take away 1), which means 4. This is an example of the second rule.

For the number 6, the Romans used VI, which means "V add I" (or 5 add 1). This is an example of the third rule.

In the same way, XL means "L take away X", or 50 − 10, which is 40.

LX means "L add X", or 50 + 10, which is 60.

Exercise 7

Change these Roman numerals into modern numbers.

a) XII   b) XX

c) XV   d) CL

e) CCX   f) MM

Here are some years shown as modern numbers. Change them into Roman numerals.

g) 2010   h) 2015

i) 1990   j) 1800

Co-ordinates

We use **co-ordinates** to give the position of something on a grid. Look at the grid below.

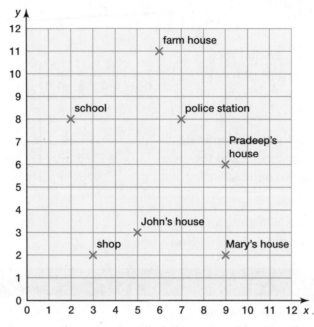

The line that runs across the page is called the *x*-axis. The line that runs up the page is called the *y*-axis.

We can show the position of something on the grid by writing the co-ordinates. For example, on our grid the position of John's house is 5 on the *x*-axis and 3 on the *y*-axis. You can see this by running your finger down from John's house to the *x*-axis and along from John's house to the *y*-axis. Your finger will reach the *x*-axis at 5 and the *y*-axis at 3.

We write this position as (5, 3). The 5 is the *x* co-ordinate and the 3 is the *y* co-ordinate. We always write a pair of co-ordinates in a bracket and we always put a comma between them. The *x* co-ordinate is always written first and then the *y* co-ordinate. You can remember this by remembering that *x* comes before *y* in the alphabet.

Exercise 8

Look at the grid above and then write the co-ordinates for each of these buildings:

a) school

b) shop

c) Mary's house

d) farm house

e) police station

f) Pradeep's house

Points can also be found on the other side of the *x*-axis and *y*-axis. We use negative numbers to show on which side of each axis the point lies.

Look at this grid.

On this grid you can see that the *x* co-ordinate for the post office is −2 and the *y* co-ordinate is 4. So the post office is at (−2, 4).

You can also see that the *x* co-ordinate for the teacher's house is −3 and the *y* co-ordinate is −4. So the teacher's house is at (−3, −4).

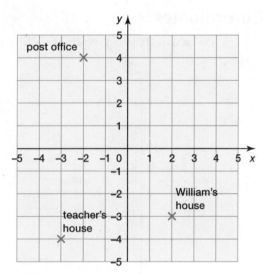

The *x* co-ordinate for William's house is 2 and the *y* co-ordinate is −3, so William's house is at (2, −3).

Exercise 9

Make a copy of the grid below.

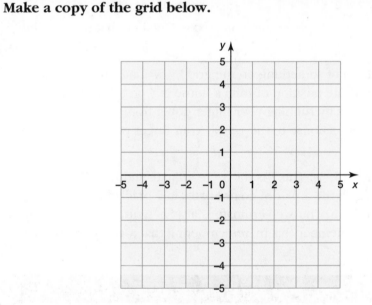

When we write a point on a grid we call it "plotting" the point on the grid. Plot these points on the grid.

a) (2, 3)

b) (−2, 3)

c) (−3, 1)

d) (−3, −2)

e) (−1, −3)

f) (3, −3)

Remember!

- The multiples of a number are all the numbers we get when we multiply the number by 1, 2, 3, 4, 5 … and so on.
- Finding the lowest common multiple of two numbers is often very useful in mathematics.
- The factors of a number are all the numbers that divide into it exactly.
- Finding the highest common factor of two numbers is often very useful in mathematics.
- Prime numbers have only two factors: the number itself and 1.
- If we multiply a number by itself we get a square number.
- There are three rules we must follow for making numbers with Roman numerals, as given on page 46.
- We use x and y co-ordinates to find points on a grid. The co-ordinates are written in a bracket with a comma between them, like this (x, y). The x co-ordinate is always written first, then the y co-ordinate.

Revision Test on More About Numbers

Now that you have worked your way through the chapter, try this revision test. The answers are in the answer book.

1. Write down the first five multiples of 7.

2. Which of these numbers are multiples of 5?
   20, 28, 35, 70, 95, 158, 305, 551

3. Write down the lowest common multiple of each of these pairs of numbers:
   a) 2 and 5   b) 3 and 4
   c) 3 and 8   d) 5 and 7
   e) 4 and 9   f) 5 and 11
   g) 6 and 8   h) 7 and 9

4. Write down all the factors of these numbers:
   a) 6   b) 7
   c) 9   d) 12
   e) 16   f) 26
   g) 40   h) 100

5. Write down the highest common factor of each of these pairs of numbers:
   a) 4 and 10   b) 9 and 18
   c) 6 and 21   d) 8 and 12
   e) 12 and 22   f) 15 and 40
   g) 12 and 32   h) 27 and 45

6. Which of these numbers are prime numbers?

   4, 7, 9, 12, 17, 19, 24, 27, 37, 41

7. What are these square numbers equal to?

   a) 4^2 b) 6^2

   c) 10^2 d) 11^2

   e) 20^2 f) 100^2

8. Give the answers to these addition and subtraction questions:

   a) $2^2 + 3^2 =$ b) $4^2 + 1^2 =$

   c) $3^2 + 5^2 =$ d) $6^2 + 2^2 =$

   e) $6^2 - 2^2 =$ f) $4^2 - 3^2 =$

   g) $10^2 - 5^2 =$ h) $7^2 - 3^2 =$

9. Change these numbers into Roman numerals:

   a) 9   b) 15

   c) 23   d) 69

   e) 300   f) 522

   g) 1066   h) 1789

10. Change these Roman numerals into modern numbers:

    a) VIII   b) XXII

    c) XL   d) CX

    e) MX   f) MMMXX

    g) CCL   h) XIX

11. Copy this grid.

    Now plot each of these points on your grid.

    a) $(1, 3)$

    b) $(2, 4)$

    c) $(-3, 4)$

    d) $(-4, -4)$

    e) $(1, -1)$

    f) $(0, 2)$

12. Look at this grid.

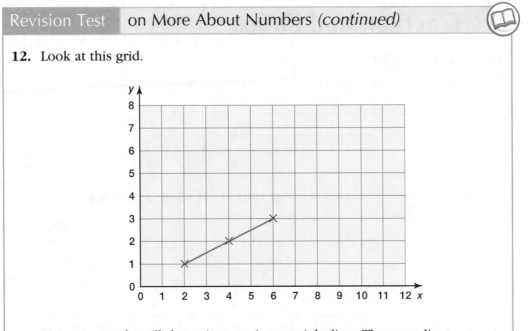

You can see that all the points are in a straight line. The co-ordinates for the three points are (2, 1), (4, 2) and (6, 3). For every point, the *x* co-ordinate is twice as big as the *y* co-ordinate. What would be the co-ordinates of the next two points on this straight line?

6 Fractions and Ratios

What is a Fraction?

If we divide a shape into equal parts, each part is called a **fraction**.

Here is a shape divided into 2 equal parts:

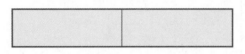

Each part is the same size. Each part is a **half** of the whole shape. A half is a fraction and we write it as:

$$\frac{1}{2}$$

If we divide the shape into 3 equal parts, it will look like this:

Each part is the same size. This time each part is called a **third** of the whole shape. A third is a fraction and we write it like this:

$$\frac{1}{3}$$

Here is a round cake divided into 4 equal pieces:

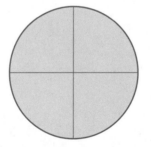

Each part is the same size. Each part is called a **quarter** of the cake and we write it as a fraction like this:

$$\frac{1}{4}$$

The number at the bottom of a fraction is called the **denominator**. This number tells us how many equal parts something has been divided into. The number on the top of a fraction is called the **numerator**. This number tells us how many parts we have.

In the example of our cake divided into 4 equal pieces, 1 piece would be written as:

$\dfrac{1}{4}$ — this is the numerator (how many pieces we have)

— this is the denominator (how many equal pieces there are altogether)

This tells us that the cake has been divided into 4 equal pieces and we have 1 piece.

If the cake is divided into 6 equal pieces and we have 1 piece, we would write it as a fraction like this:

$$\dfrac{1}{6}$$

Exercise 1

Here are some fractions of a shape. For each fraction, write down how many equal parts the shape has been divided into.

a) $\dfrac{1}{3}$ b) $\dfrac{1}{4}$

c) $\dfrac{1}{5}$ d) $\dfrac{1}{12}$

e) $\dfrac{1}{2}$ f) $\dfrac{1}{8}$

g) $\dfrac{1}{10}$ h) $\dfrac{1}{15}$

You can see from your answers to the exercise above that the bigger the number on the bottom of a fraction, the more parts the shape has been divided into. Let's compare $\dfrac{1}{3}$ with $\dfrac{1}{6}$.

In this example, the shape is divided into 3 equal parts and each part is $\dfrac{1}{3}$:

In this example, the shape is divided into 6 equal parts and each part is $\dfrac{1}{6}$:

We can see from these shapes that $\dfrac{1}{3}$ is bigger than $\dfrac{1}{6}$.

If you look carefully at the shapes you will see that $\frac{1}{3}$ is twice as big as $\frac{1}{6}$.

How Big are Fractions?

We can show the size of fractions by drawing a **fraction wall**. The wall is the same width at the top and the bottom, but each row in the wall is divided into different-sized fractions.

$\frac{1}{2}$						$\frac{1}{2}$					
$\frac{1}{3}$				$\frac{1}{3}$				$\frac{1}{3}$			
$\frac{1}{4}$			$\frac{1}{4}$			$\frac{1}{4}$			$\frac{1}{4}$		
$\frac{1}{5}$		$\frac{1}{5}$		$\frac{1}{5}$		$\frac{1}{5}$		$\frac{1}{5}$			
$\frac{1}{6}$		$\frac{1}{6}$		$\frac{1}{6}$		$\frac{1}{6}$		$\frac{1}{6}$		$\frac{1}{6}$	
$\frac{1}{7}$	$\frac{1}{7}$	$\frac{1}{7}$	$\frac{1}{7}$	$\frac{1}{7}$	$\frac{1}{7}$	$\frac{1}{7}$					
$\frac{1}{8}$	$\frac{1}{8}$	$\frac{1}{8}$	$\frac{1}{8}$	$\frac{1}{8}$	$\frac{1}{8}$	$\frac{1}{8}$	$\frac{1}{8}$				
$\frac{1}{9}$	$\frac{1}{9}$	$\frac{1}{9}$	$\frac{1}{9}$	$\frac{1}{9}$	$\frac{1}{9}$	$\frac{1}{9}$	$\frac{1}{9}$	$\frac{1}{9}$			
$\frac{1}{10}$	$\frac{1}{10}$	$\frac{1}{10}$	$\frac{1}{10}$	$\frac{1}{10}$	$\frac{1}{10}$	$\frac{1}{10}$	$\frac{1}{10}$	$\frac{1}{10}$	$\frac{1}{10}$		
$\frac{1}{11}$	$\frac{1}{11}$	$\frac{1}{11}$	$\frac{1}{11}$	$\frac{1}{11}$	$\frac{1}{11}$	$\frac{1}{11}$	$\frac{1}{11}$	$\frac{1}{11}$	$\frac{1}{11}$	$\frac{1}{11}$	
$\frac{1}{12}$	$\frac{1}{12}$	$\frac{1}{12}$	$\frac{1}{12}$	$\frac{1}{12}$	$\frac{1}{12}$	$\frac{1}{12}$	$\frac{1}{12}$	$\frac{1}{12}$	$\frac{1}{12}$	$\frac{1}{12}$	$\frac{1}{12}$

Spend a few minutes looking at the fraction wall. You can see that fractions get smaller as the bottom number (the denominator) gets bigger. For example, $\frac{1}{12}$ is much smaller than $\frac{1}{3}$.

You can also see that if you add 2 quarters together, they are the same size as 1 half.

We can write $\frac{1}{4} + \frac{1}{4}$ as $\frac{2}{4}$.

This means we have divided the shape into 4 equal parts (the bottom number) and we now have 2 of the parts (the top number).

We can see from the fraction wall that $\frac{1}{4} + \frac{1}{4}$ is the same as $\frac{2}{4}$ and this is the same as $\frac{1}{2}$. So:

$$\frac{2}{4} = \frac{1}{2}$$

So 2 quarters are equal to 1 half. When two fractions are the same size we call them **equivalent fractions**.

By looking at the fraction wall you can find some other equivalent fractions. Here are a few others:

$$\frac{3}{6} = \frac{1}{2} \qquad \frac{1}{4} = \frac{2}{8} \qquad \frac{3}{4} = \frac{9}{12}$$

There are many more. How many can you find on the wall?

How Do We Know If Fractions Are the Same Size?

If we don't have a fraction wall or a shape to look at, we need to find another way of working out whether two fractions are the same size.

We do this by multiplying or dividing the top number and the bottom number of a fraction by the same number. The answer we get will be a fraction of the same size.

For example, if we take $\frac{1}{2}$ and multiply the top number and the bottom number by 2 we will get $\frac{2}{4}$. If we multiply the top number and the bottom number again by 2 we will get $\frac{4}{8}$. So this tells us that $\frac{1}{2}$, $\frac{2}{4}$ and $\frac{4}{8}$ are all the same size. They are equivalent fractions.

If you think about this, it is obvious. $\frac{2}{4}$ means we have a shape divided into 4 equal pieces and we have 2 of them, which is half of all the pieces. Remember that we write half as $\frac{1}{2}$.

$\frac{4}{8}$ means we have a shape divided into 8 equal pieces and we have 4 of them, which is also half of all the pieces.

So $\frac{2}{4}$, $\frac{4}{8}$ and $\frac{1}{2}$ are the same size.

Below is an example where we divide the top number and the bottom number by the same number. When we are dividing, we have to find a number that will divide into both the top number and the bottom number.

Let's take $\frac{6}{9}$ as our example. We need to find a number that divides into 6 and into 9. If we think about this for a few moments we can see that 3 divides into both 6 and 9. So if we divide the top number and the bottom number by 3 we will get $\frac{2}{3}$. This tells us that $\frac{6}{9}$ is the same size as $\frac{2}{3}$:

$$\frac{6}{9} = \frac{2}{3}$$

They are equivalent fractions.

The equivalent fraction with the smallest bottom number is the fraction in its **simplest form**. Fractions in the simplest form are often called fractions in their **lowest terms**.

$\frac{8}{16}$, $\frac{4}{8}$ and $\frac{1}{2}$ are all the same size, but in its simplest form or lowest terms this fraction is:

$\frac{1}{2}$

Exercise 2

Look at this list of fractions.

$$\frac{2}{4} \qquad \frac{6}{8} \qquad \frac{2}{6} \qquad \frac{6}{10}$$

a) Which fraction is the same size as $\frac{1}{3}$?

b) Which fraction is the same size as $\frac{1}{2}$?

c) Which fraction is the same size as $\frac{3}{4}$?

d) Which fraction is the same size as $\frac{3}{5}$?

Write the following fractions in their lowest terms.

e) $\dfrac{2}{4}$ f) $\dfrac{3}{6}$

g) $\dfrac{4}{6}$ h) $\dfrac{5}{10}$

i) $\dfrac{6}{8}$ j) $\dfrac{3}{12}$

For each of these pairs of equivalent fractions, find the missing number.

k) $\dfrac{1}{2} = \dfrac{\square}{8}$ l) $\dfrac{3}{4} = \dfrac{\square}{12}$

m) $\dfrac{1}{2} = \dfrac{3}{\square}$ n) $\dfrac{1}{3} = \dfrac{\square}{9}$

o) $\dfrac{\square}{5} = \dfrac{2}{10}$ p) $\dfrac{80}{100} = \dfrac{\square}{10}$

Comparing and Ordering Fractions

We have already seen in the fraction wall that fractions get smaller as the bottom number gets bigger. So $\frac{1}{12}$ is smaller than $\frac{1}{3}$.

If the bottom number is the same, we can see which fraction is bigger by looking at the top number. So if we are comparing $\frac{3}{5}$ and $\frac{2}{5}$ we can see that $\frac{3}{5}$ is the biggest.

If we think of a shape divided into 5 equal pieces, we can see that 3 pieces will be bigger than 2 pieces, so:

$\frac{3}{5} > \frac{2}{5}$

But how can we tell which fraction is bigger if the top *and* bottom numbers are different?

For example, which is bigger, $\frac{2}{3}$ or $\frac{3}{4}$?

To work this out without looking at the fraction wall we need to make the bottom number the same in both fractions. To do this we need to find a common multiple of the bottom numbers. The quickest way to do this is to multiply the top and bottom number of each fraction by the bottom number of the other fraction.

So if we multiply the top and bottom of $\frac{2}{3}$ by 4 (the bottom number of the second fraction) we get $\frac{8}{12}$. If we multiply the top and bottom of $\frac{3}{4}$ by 3 (the bottom number of the first fraction) we will get $\frac{9}{12}$.

12 is the lowest common multiple of 3 and 4. If you have forgotten how to work out common multiples, look back at Chapter 5 *More About Numbers*.

So now, instead of comparing $\frac{2}{3}$ and $\frac{3}{4}$ to see which is biggest, we are comparing the fractions $\frac{8}{12}$ and $\frac{9}{12}$. We can see that $\frac{9}{12}$ is bigger than $\frac{8}{12}$. If you can't see this, think about a shape divided into 12 equal pieces. $\frac{9}{12}$ means 9 pieces of the shape. $\frac{8}{12}$ means 8 pieces of the shape. So $\frac{9}{12}$ is bigger:

$\frac{9}{12} > \frac{8}{12}$

So $\frac{3}{4} > \frac{2}{3}$.

Exercise 3

Put the fractions into size order giving the biggest first.

a) $\frac{1}{2}$ and $\frac{1}{3}$

b) $\frac{1}{2}$ and $\frac{3}{5}$

c) $\frac{1}{3}$ and $\frac{1}{4}$

d) $\frac{2}{3}$ and $\frac{5}{9}$

e) $\frac{3}{4}$ and $\frac{5}{6}$

f) $\frac{2}{3}$ and $\frac{4}{5}$

g) $\frac{5}{16}$ and $\frac{3}{8}$ and $\frac{1}{4}$

h) $\frac{2}{3}$ and $\frac{9}{15}$ and $\frac{4}{5}$

Adding and Subtracting Fractions

If we want to add two fractions together we must first make sure that the bottom number is the same. For example, if we want to add $\frac{1}{3} + \frac{1}{3}$ it is easy as the bottom number is already the same. We are just adding 1 third to another 1 third, which will give us 2 thirds. We write this as $\frac{2}{3}$:

$$\frac{1}{3} + \frac{1}{3} = \frac{2}{3}$$

Here is another example. Adding one-fifth to three-fifths gives four-fifths:

$$\frac{1}{5} + \frac{3}{5} = \frac{4}{5}$$

When adding fractions with the same bottom number we just add the top numbers together.

Exercise 4

Add together the following fractions.

a) $\frac{1}{5} + \frac{1}{5} =$ b) $\frac{1}{7} + \frac{1}{7} =$

c) $\frac{1}{5} + \frac{2}{5} =$ d) $\frac{2}{7} + \frac{2}{7} =$

e) $\frac{1}{9} + \frac{4}{9} =$ f) $\frac{2}{9} + \frac{5}{9} =$

If we want to add together fractions with different bottom numbers, such as $\frac{1}{3} + \frac{1}{2}$, we must first change the bottom numbers so that they are the same.

To make the bottom number the same we must multiply the top and bottom numbers of each fraction by numbers that will give us the same bottom number. Remember that the bottom number of a fraction is called the denominator, so when two fractions have the same bottom number we call this a **common denominator**.

So for our example:

$$\frac{1}{3} + \frac{1}{2} =$$

we need to find a common denominator. If we multiply the top and bottom of the first fraction by the bottom number of the second fraction we will get $\frac{1}{3}$ multiplied top and bottom by 2 to give us $\frac{2}{6}$, and if we multiply the top and bottom of $\frac{1}{2}$ by the bottom number of the first fraction we will get $\frac{3}{6}$.

So now we have:

$$\frac{2}{6} + \frac{3}{6} =$$

Now we have a common denominator so we can add the fractions together:

$$\frac{2}{6} + \frac{3}{6} = \frac{5}{6}$$

So now we can write out our example in full:

$$\frac{1}{3} + \frac{1}{2} = \frac{2}{6} + \frac{3}{6} = \frac{5}{6}$$

Here is another example:

$$\frac{1}{5} + \frac{4}{10} =$$

We could multiply the top and bottom of $\frac{1}{5}$ by 10 and the top and bottom of $\frac{4}{10}$ by 5 to give us $\frac{10}{50}$ and $\frac{20}{50}$.

But a quicker way is to spot that 10 is the lowest common multiple of 5 and 10.

So if we multiply the top and bottom of $\frac{1}{5}$ by 2 to give $\frac{2}{10}$ we would then have:

$$\frac{2}{10} + \frac{4}{10} = \frac{6}{10}$$

$\frac{6}{10}$ is the same size as $\frac{3}{5}$. They are equivalent fractions. Normally in mathematics we give the equivalent fraction with the smallest denominator as the answer, so if we have a choice of writing $\frac{6}{10}$ or $\frac{3}{5}$ we would normally write $\frac{3}{5}$ as this is the fraction in its lowest terms.

Another way to work out $\frac{1}{5} + \frac{4}{10}$ is to divide the top and bottom of $\frac{4}{10}$ by 2 to get a common denominator. This will give us:

$$\frac{1}{5} + \frac{2}{5} = \frac{3}{5}$$

Exercise 5

Add together the following fractions. Write the answers in the lowest terms.

a) $\frac{2}{3} + \frac{1}{6} =$

b) $\frac{1}{2} + \frac{1}{4} =$

c) $\frac{1}{3} + \frac{1}{9} =$

d) $\frac{1}{5} + \frac{1}{15} =$

Exercise 5 *(continued)*

e) $\dfrac{3}{4} + \dfrac{1}{8} =$

f) $\dfrac{1}{4} + \dfrac{1}{3} =$

g) $\dfrac{1}{2} + \dfrac{2}{5} =$

h) $\dfrac{1}{6} + \dfrac{2}{9} =$

Subtracting fractions is very like adding fractions. We need to find a common bottom number and then subtract the top numbers. Here are two examples:

$$\frac{2}{3} - \frac{1}{3} = \frac{1}{3} \qquad \frac{3}{4} - \frac{1}{4} = \frac{2}{4} \text{ which equals } \frac{1}{2}$$

If the bottom numbers are different we need to find a common bottom number to make the subtraction work, in the same way we do for adding fractions. For example:

$$\frac{5}{6} - \frac{2}{3} \text{ is the same as } \frac{5}{6} - \frac{4}{6} \text{ which equals } \frac{1}{6}$$

This works because $\frac{2}{3}$ and $\frac{4}{6}$ are the same. They are equivalent fractions.

Here is another example:

$$\frac{2}{3} - \frac{1}{4} =$$

The lowest common multiple of 3 and 4 is 12. We multiply the top and bottom of $\frac{2}{3}$ by 4 to get $\frac{8}{12}$. We then multiply the top and bottom of $\frac{1}{4}$ by 3 to get $\frac{3}{12}$. Now we have:

$$\frac{2}{3} - \frac{1}{4} = \frac{8}{12} - \frac{3}{12} = \frac{5}{12}$$

So the answer is $\frac{5}{12}$.

Here is one more example:

$$\frac{5}{6} - \frac{3}{5} =$$

The lowest common multiple of 6 and 5 is 30. We multiply the top and bottom of each fraction by the bottom number of the other fraction to get to 30 as the common denominator. So the answer is:

$$\frac{5}{6} - \frac{3}{5} = \frac{25}{30} - \frac{18}{30} = \frac{7}{30}$$

Exercise 6

Subtract the following fractions. Write the answers in the lowest terms.

a) $\dfrac{4}{5} - \dfrac{1}{5} =$

b) $\dfrac{5}{6} - \dfrac{1}{6} =$

c) $\dfrac{14}{15} - \dfrac{11}{15} =$

d) $\dfrac{2}{3} - \dfrac{1}{9} =$

e) $\dfrac{6}{7} - \dfrac{1}{14} =$

f) $\dfrac{2}{3} - \dfrac{1}{2} =$

g) $\dfrac{2}{3} - \dfrac{2}{5} =$

h) $\dfrac{4}{7} - \dfrac{1}{4} =$

Rules for adding and subtracting fractions

1. Find a common denominator for the fractions.
2. Write each fraction as an equivalent fraction with the common denominator.
3. Add or subtract the numerators of the new fractions.
4. Write the answer in its lowest terms. For example, $\frac{2}{4}$ should be written as $\frac{1}{2}$.

Types of Fractions

Proper Fractions

All the fractions we have looked at so far have a top number *smaller than* the bottom number. These fractions are called **proper fractions**. Here are three examples of proper fractions:

$$\frac{1}{3}, \frac{5}{6}, \frac{3}{10}$$

Improper Fractions

Fractions that have a top number *bigger than or equal to* the bottom number are called **improper fractions**. Here are three examples of improper fractions:

$$\frac{3}{2}, \frac{5}{3}, \frac{6}{6}$$

Improper fractions where the top and bottom numbers are equal are *always* equal to 1. For example:

$$\frac{6}{6} = 1$$

Remember that the bottom number tells us how many equal pieces something has been divided into, and the top number tells us how many pieces we have. Think about a circle:

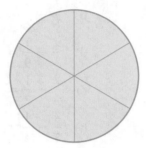

$\frac{6}{6}$ tells us that the circle has been divided into 6 equal pieces and we have all 6 pieces. So we have 1 whole circle. So $\frac{6}{6} = 1$.

Here are some more examples of improper fractions that equal 1. They all have equal top and bottom numbers:

$$\frac{3}{3} = 1 \qquad \frac{5}{5} = 1 \qquad \frac{10}{10} = 1 \qquad \frac{15}{15} = 1 \qquad \frac{65}{65} = 1$$

Improper fractions with a top number bigger than the bottom number are greater than 1.

Mixed Numbers

A whole number and a proper fraction together are called a **mixed number**. Here are three examples of mixed numbers:

$$1\frac{3}{5}, 3\frac{1}{3}, 5\frac{6}{7}$$

We can think of a mixed number as a whole number and a fraction added together. For example:

$$3\frac{1}{3} = 3 + \frac{1}{3}$$

$$5\frac{6}{7} = 5 + \frac{6}{7}$$

Changing Improper Fractions into Mixed Numbers

Sometimes if we have an improper fraction like $\frac{5}{4}$ we want to change it into a mixed number. We know that:

$$\frac{5}{4} = \frac{4}{4} + \frac{1}{4}$$

We also know that $\frac{4}{4} = 1$, so we can replace $\frac{4}{4}$ with 1 to give us:

$$\frac{5}{4} = 1 + \frac{1}{4} = 1\frac{1}{4}$$

The quickest way to change an improper fraction into a mixed number is to divide the top number by the bottom number. This gives us the whole number, and the remainder is the top number of the fraction left over. The fraction has the same bottom number as the improper fraction we started with.

So for $\frac{5}{4}$ divide the top number by the bottom number:

$$5 \div 4 = 1 \text{ r1}$$

The remainder means that there is 1 quarter left over. So the mixed number is $1\frac{1}{4}$.

Let us write $\frac{8}{3}$ as a mixed number. First we divide the top number by the bottom number. That gives us the whole number, and any remainder is the top number of the fraction left over:

$$8 \div 3 = 2 \text{ r2}$$

The remainder means there are 2 thirds left over. So the mixed number is $2\frac{2}{3}$.

We always write the fraction part of a mixed number in its lowest terms.

Exercise 7

Change these improper fractions into mixed numbers by dividing the top number by the bottom number.

a) $\dfrac{3}{2}$

b) $\dfrac{4}{3}$

c) $\dfrac{8}{3}$

d) $\dfrac{10}{3}$

e) $\dfrac{9}{2}$

f) $\dfrac{11}{5}$

g) $\dfrac{15}{6}$

h) $\dfrac{32}{10}$

Changing Mixed Numbers into Improper Fractions

We can change mixed numbers into improper fractions by remembering that any fraction where the top number and the bottom number are the same is equal to 1.

So if we have $1\frac{1}{3}$ and we want to change it into an improper fraction, we can remember that $1 = \frac{3}{3}$:

$$1\frac{1}{3} = \frac{3}{3} + \frac{1}{3} = \frac{4}{3}$$

So $1\frac{1}{3} = \frac{4}{3}$.

If our mixed number is $2\frac{1}{3}$ this is the same as $1 + 1 + \frac{1}{3}$.

This is the same as:

$$\frac{3}{3} + \frac{3}{3} + \frac{1}{3} = \frac{7}{3}$$

So $2\frac{1}{3} = \frac{7}{3}$.

The quickest way to change a mixed number into an improper fraction is to multiply the whole number by the bottom number of the fraction and then add the top number of the fraction to give us the new top number. The bottom number in the improper fraction we end with is the same as in the mixed number we started with.

So if we have $2\frac{2}{3}$ and we want to change it to an improper fraction we first multiply 2 by the bottom number of the fraction:

$$2 \times 3 = 6$$

Then we add the top number of the fraction to give the new top number:

$$6 + 2 = 8$$

The bottom number stays the same, so the improper fraction is $\frac{8}{3}$:

$$2\frac{2}{3} = \frac{8}{3}$$

Here is one more example. Let's change $3\frac{2}{3}$ into an improper fraction:

$$3 \times 3 = 9$$

$$9 + 2 = 11$$

So the answer is:

$$3\frac{2}{3} = \frac{11}{3}$$

Exercise 8

Change these mixed numbers into improper fractions.

a) $1\frac{1}{2}$

b) $1\frac{4}{5}$

c) $2\frac{1}{6}$

d) $2\frac{3}{4}$

e) $3\frac{3}{5}$

f) $5\frac{1}{3}$

g) $2\frac{7}{8}$

h) $10\frac{1}{2}$

Adding Mixed Numbers

There are two ways of adding mixed numbers. Try both ways to see which works best for you.

The first way is to change the mixed numbers into improper fractions and then add them together.

So for $1\frac{1}{2} + 3\frac{3}{4}$ we can use the method we have learnt to turn the mixed numbers into improper fractions:

$$1\frac{1}{2} + 3\frac{3}{4} \text{ is the same as } \frac{3}{2} + \frac{15}{4}$$

Now we need to find a common bottom number. The lowest common multiple of the bottom numbers is 4, so we multiply the top and bottom of $\frac{3}{2}$ by 2 to give $\frac{6}{4}$. Now we have:

$$\frac{6}{4} + \frac{15}{4} = \frac{21}{4}$$

We can turn this answer back into a mixed number by dividing the bottom number into the top number to give 5 r1, which gives us $5\frac{1}{4}$.

So if we put it all together we get:

$$1\frac{1}{2} + 3\frac{3}{4} = \frac{3}{2} + \frac{15}{4} = \frac{6}{4} + \frac{15}{4} = \frac{21}{4} = 5\frac{1}{4}$$

The second way to add mixed numbers is to add the whole number parts together first and then add the fraction parts separately. So for our example:

$$1\frac{1}{2} + 3\frac{3}{4} =$$

First let's add the whole numbers:

$$1 + 3 = 4$$

Now we add the fractions:

$$\frac{1}{2} + \frac{3}{4} =$$

Now we need to find a common bottom number. The lowest common multiple is 4, which will give us:

$$\frac{2}{4} + \frac{3}{4} = \frac{5}{4} = 1\frac{1}{4}$$

Now we need to add the whole number answer and the fraction answer together:

$$4 + 1\frac{1}{4} = 5\frac{1}{4}$$

Exercise 9

Add the following mixed numbers using the method that works best for you.

a) $1\frac{1}{4} + 1\frac{1}{8} =$ b) $3\frac{1}{2} + 2\frac{1}{4} =$

c) $2\frac{3}{4} + 1\frac{1}{2} =$ d) $2\frac{1}{3} + 3\frac{1}{6} =$

e) $2\frac{1}{3} + 1\frac{1}{9} =$ f) $1\frac{1}{10} + 2\frac{1}{5} =$

Subtracting Mixed Numbers

We can subtract mixed numbers in the same way that we add them. We just subtract instead of add!

So if we want to work out $3\frac{3}{4} - 1\frac{1}{2}$ we can use either of our two methods:

1. Change the mixed numbers into improper fractions and then subtract.

$$3\frac{3}{4} - 1\frac{1}{2} = \frac{15}{4} - \frac{3}{2} = \frac{15}{4} - \frac{6}{4} = \frac{9}{4} = 2\frac{1}{4}$$

2. **Subtract the whole number parts first, then subtract the fraction parts, and then add the two parts together.**

$$3 - 1 = 2$$

$$\frac{3}{4} - \frac{1}{2} = \frac{3}{4} - \frac{2}{4} = \frac{1}{4}$$

then we add the two parts together $2 + \frac{1}{4} = 2\frac{1}{4}$

So $3\frac{3}{4} - 1\frac{1}{2} = 2\frac{1}{4}$.

Most people find it easier to use the first method. Try both and see which works best for you.

Exercise 10

Subtract these mixed numbers.

a) $3\frac{1}{2} - 1\frac{1}{4} =$

b) $3\frac{1}{4} - 1\frac{1}{2} =$

c) $2\frac{2}{3} - 1\frac{1}{3} =$

d) $4\frac{1}{2} - 3\frac{1}{4} =$

e) $2\frac{2}{3} - 1\frac{1}{6} =$

f) $5\frac{1}{4} - 3\frac{3}{4} =$

Multiplying Fractions

If we want to multiply two fractions together we first multiply the top numbers, then we multiply the bottom numbers to give us the answer.

For example:

$$\frac{2}{3} \times \frac{1}{4} = \frac{2 \times 1}{3 \times 4} = \frac{2}{12}$$

We can reduce $\frac{2}{12}$ to its lowest terms, and write it as $\frac{1}{6}$.

Here is another example:

$$\frac{4}{5} \times \frac{5}{6} = \frac{4 \times 5}{5 \times 6} = \frac{20}{30} = \frac{2}{3}$$

If a question has a mixed number, we always convert it into an improper fraction first before multiplying. For example:

$$2\frac{1}{2} \times \frac{1}{4} = \frac{5}{2} \times \frac{1}{4}$$

Now we multiply the fractions together by multiplying the top numbers and the bottom numbers:

$$\frac{5}{2} \times \frac{1}{4} = \frac{5 \times 1}{2 \times 4} = \frac{5}{8}$$

If we want to multiply a fraction by a whole number we use the same method. Remember that any whole number written as a fraction is just the whole number over 1. For example, 2 is the same as $\frac{2}{1}$ and 3 is the same as $\frac{3}{1}$ and so on.

So if we want to find the answer to $3 \times \frac{1}{2}$ we can write this as $\frac{3}{1} \times \frac{1}{2}$:

$$3 \times \frac{1}{2} = \frac{3}{1} \times \frac{1}{2} = \frac{3 \times 1}{1 \times 2} = \frac{3}{2} = 1\frac{1}{2}$$

Here is another example. Let's find the answer to $5 \times \frac{2}{3}$:

$$5 \times \frac{2}{3} = \frac{5}{1} \times \frac{2}{3} = \frac{5 \times 2}{1 \times 3} = \frac{10}{3} = 3\frac{1}{3}$$

Exercise 11

Multiply these fractions. Give the answers in their lowest terms.

a) $2 \times \frac{3}{4} =$

b) $3 \times \frac{3}{5} =$

c) $4 \times \frac{1}{3} =$

d) $6 \times \frac{2}{5} =$

e) $\frac{1}{2} \times \frac{1}{4} =$

f) $\frac{4}{5} \times \frac{2}{3} =$

g) $\frac{5}{6} \times \frac{3}{4} =$

h) $\frac{1}{10} \times \frac{2}{3} =$

i) $2\frac{1}{5} \times \frac{1}{2} =$

j) $1\frac{1}{2} \times 2\frac{1}{3} =$

k) $\frac{3}{4} \times 2\frac{3}{4} =$

l) $6\frac{2}{3} \times 2\frac{1}{8} =$

Finding a Fraction of an Amount

If we want to find a fraction of an amount we multiply the amount by the fraction we need. For example, if we want to know what $\frac{2}{3}$ of 18 is, we multiply 18 by $\frac{2}{3}$:

$$18 \times \frac{2}{3} = \frac{18}{1} \times \frac{2}{3} = \frac{18 \times 2}{1 \times 3} = \frac{36}{3} = 12$$

We can use this method to solve word problems like this one:

"There are 24 biscuits in a shop. Eleanor buys $\frac{2}{3}$ of the biscuits. How many biscuits does she buy?"

We need to work out what $\frac{2}{3}$ of 24 is. We can do this by multiplying 24 by $\frac{2}{3}$:

$$24 \times \frac{2}{3} = \frac{24}{1} \times \frac{2}{3} = \frac{24 \times 2}{1 \times 3} = \frac{48}{3} = 16$$

So Eleanor buys 16 biscuits.

Here is another word problem:

> "There are 30 pupils in a class. $\frac{3}{5}$ say that football is their favourite sport. How many pupils say that football is their favourite sport?"

We need to find out what $\frac{3}{5}$ of 30 is, so we multiply 30 by $\frac{3}{5}$:

$$30 \times \frac{3}{5} = \frac{30}{1} \times \frac{3}{5} = \frac{30 \times 3}{1 \times 5} = \frac{90}{5} = 18$$

So 18 pupils say that football is their favourite sport.

Dividing Fractions

If we want to divide a whole number by a fraction, we turn the fraction upside down and then multiply. For example, $4 \div \frac{1}{2}$ is the same as $4 \times \frac{2}{1}$:

$$4 \div \frac{1}{2} = 4 \times \frac{2}{1} = \frac{4}{1} \times \frac{2}{1} = \frac{4 \times 2}{1 \times 1} = \frac{8}{1} = 8$$

So $4 \div \frac{1}{2} = 8$.

If we want to divide a fraction by another fraction, we turn the second fraction upside down and multiply. For example:

$$\frac{1}{3} \div \frac{1}{6} = \frac{1}{3} \times \frac{6}{1} = \frac{1 \times 6}{3 \times 1} = \frac{6}{3} = 2$$

If we want to divide mixed numbers we first turn them into improper fractions. So if we want to work out $3\frac{1}{2} \div \frac{3}{4}$ we change it into $\frac{7}{2} \div \frac{3}{4}$:

$$3\frac{1}{2} \div \frac{3}{4} = \frac{7}{2} \div \frac{3}{4}$$

Next we turn the second fraction upside down and multiply:

$$\frac{7}{2} \div \frac{3}{4} = \frac{7}{2} \times \frac{4}{3} = \frac{7 \times 4}{2 \times 3} = \frac{28}{6} = 4\frac{4}{6} = 4\frac{2}{3}$$

Here is another example:

$$5\frac{3}{4} \div 2\frac{2}{3} = \frac{23}{4} \div \frac{8}{3} = \frac{23}{4} \times \frac{3}{8} = \frac{23 \times 3}{4 \times 8} = \frac{69}{32} = 2\frac{5}{32}$$

Here is a word problem that we can answer by dividing fractions:

> "Ruth has three drawers in her desk. Each of her school books fills $\frac{1}{5}$ of a drawer. How many school books can she fit into her drawers altogether?"

Each book fills $\frac{1}{5}$ of a drawer so if we know how many fifths there are in 3 drawers we will know how many books will fit. To work this out we need to divide 3 by $\frac{1}{5}$:

$$3 \div \frac{1}{5} = \frac{3}{1} \div \frac{1}{5} = \frac{3}{1} \times \frac{5}{1} = \frac{3 \times 5}{1 \times 1} = \frac{15}{1} = 15$$

So 15 books will fit in the drawers altogether.

Exercise 12

Give answers to these questions in their lowest terms.

a) $\dfrac{1}{2} \div \dfrac{1}{4} =$

b) $6 \div \dfrac{1}{3} =$

c) $\dfrac{3}{15} \div \dfrac{4}{5} =$

d) $1\dfrac{1}{2} \div 2\dfrac{3}{4} =$

e) $3\dfrac{2}{3} \div \dfrac{8}{9} =$

f) $1\dfrac{1}{2} \div 1\dfrac{3}{4} =$

Names of Fractions

Here is a table that gives the names of some of the fractions. You will know many of them already.

Fraction	Name	Fraction	Name
$\frac{1}{2}$	one-half	$\frac{1}{9}$	one-ninth
$\frac{1}{3}$	one-third	$\frac{1}{10}$	one-tenth
$\frac{1}{4}$	one-quarter	$\frac{1}{20}$	one-twentieth
$\frac{1}{5}$	one-fifth	$\frac{1}{30}$	one-thirtieth
$\frac{1}{6}$	one-sixth	$\frac{1}{50}$	one-fiftieth
$\frac{1}{7}$	one-seventh	$\frac{1}{100}$	one-hundredth
$\frac{1}{8}$	one-eighth	$\frac{1}{1000}$	one-thousandth

Here are some examples of fraction names with top numbers greater than 1:

Fraction	Name	Fraction	Name
$\frac{2}{3}$	two-thirds	$\frac{3}{10}$	three-tenths
$\frac{3}{4}$	three-quarters	$\frac{31}{100}$	thirty-one-hundredths
$\frac{5}{7}$	five-sevenths	$\frac{17}{1000}$	seventeen-thousandths

Ratios

Ratios are very similar to fractions and tell us how many of one thing we have compared to another. For example, if Jack has 6 red pencils and 4 green pencils, that means the **ratio** of red pencils to green pencils is 6 to 4. This can be written as $6:4$.

Ratios can be simplified like fractions, so we can simplify Jack's ratio by dividing both sides by 2:

$6:4 = 3:2$

This tells us that for every 3 red pencils, Jack has 2 green pencils.

Ratios can be equivalent in the same way that we have equivalent fractions:

$3:2 = 6:4 = 12:8 = 24:16$

All these ratios are equivalent, but $3:2$ is the simplest form.

Here is another example. Helen counted traffic going past her school. In one hour she counted 20 cars and 5 trucks. So the ratio of cars to trucks was:

$20:5$

We can divide both numbers by 5 (5 is the highest common factor of 20 and 5):

$4:1$

So the ratio of cars to trucks driving past Helen's school in one hour was $4:1$.

Sometimes we are given a ratio and then we have to find out a number. For example:

"The ratio of teachers to pupils on a school trip was $1:10$. If five teachers went on the school trip how many pupils were there?"

The ratio tells us that for every 1 teacher there were 10 pupils. So if there were 5 teachers on the trip there must have been 5×10 pupils:

$5 \times 10 = 50$

So there were 50 pupils on the school trip.

Let's look at another example:

"Harry and Paul divided 36 marbles between them in the ratio 5:7. How many marbles did they each get?"

In this example, we add the two numbers in the ratio together to see how many equal parts, or shares, there are in the ratio altogether:

$5 + 7 = 12$

So there are 12 equal shares in the ratio. There are 36 marbles, so one share is worth $36 \div 12 = 3$ marbles.

The ratio tells us that Harry gets 5 shares and Paul gets 7 shares.

So Harry gets $5 \times 3 = 15$ marbles.

Paul gets $7 \times 3 = 21$ marbles.

Exercise 13

Write these ratios in their simplest form.

a) 2:4   b) 9:3

c) 4:12   d) 8:24

e) 6:8   f) 15:5

Find the missing numbers in these sets of equivalent ratios.

g) $3:5 = 6:$_____ $= 12:$_____

h) $8:3 = 16:$_____ $=$_____$:12$

Work out the answers to these word problems.

i)   Tom bought three apples and 12 bananas in a shop. Write the ratio of apples to bananas in its simplest form.

j)   The ratio of boys to girls in a class is 3:4. If there are 15 boys in the class, how many girls are there? How many pupils are there in the class altogether?

Remember!

◆ If we divide something into equal parts, each part is called a fraction.

◆ The number at the bottom of a fraction is called the denominator. This number tells us how many equal parts something has been divided into. The number on the top of a fraction is called the numerator. This number tells us how many parts we have.

◆ When two fractions are the same size we call them equivalent fractions.

Remember! *(continued)*

◆ To add two fractions we first make the bottom numbers the same and then we add the top numbers.

◆ To subtract two fractions we first make the bottom numbers the same and then we subtract the top numbers.

◆ To multiply two fractions we first multiply the top numbers, then we multiply the bottom numbers to give us the answer.

◆ To divide by a fraction, we turn the fraction upside down and then multiply.

◆ Ratios tell us how many of one thing we have compared to another.

Revision Test | on Fractions and Ratios

Now that you have worked your way through the chapter, try this revision test. The answers are in the answer book.

1. Which of these fractions are equivalent?

$$\frac{3}{4} \qquad \frac{2}{3} \qquad \frac{5}{6} \qquad \frac{4}{6}$$

2. Write each fraction from the first line next to the equivalent fraction in the second line.

$$\frac{10}{12} \qquad \frac{2}{3} \qquad \frac{1}{5} \qquad \frac{4}{8}$$

$$\frac{2}{10} \qquad \frac{8}{12} \qquad \frac{1}{2} \qquad \frac{5}{6}$$

3. Write each of these fractions in its lowest terms, or simplest form.

   a) $\dfrac{2}{4}$ b) $\dfrac{8}{10}$

   c) $\dfrac{4}{12}$ d) $\dfrac{5}{15}$

   e) $\dfrac{18}{27}$

4. Fill in the missing numbers in these pairs of equivalent fractions.

   a) $\dfrac{1}{3} = \dfrac{\square}{9}$ b) $\dfrac{1}{4} = \dfrac{\square}{12}$

   c) $\dfrac{2}{3} = \dfrac{\square}{21}$ d) $\dfrac{6}{\square} = \dfrac{2}{3}$

5. Put these fractions into size order. Give the smallest first.

$$\frac{1}{2} \qquad \frac{1}{4} \qquad \frac{5}{16} \qquad \frac{1}{8}$$

6. Which fraction is the largest? $\dfrac{7}{8}$ or $\dfrac{29}{32}$

$\rightarrow$

7. $\dfrac{2}{3} > \dfrac{18}{30}$ True or false?

8. $\dfrac{4}{5} > \dfrac{19}{25}$ True or false?

9. Change these improper fractions into mixed numbers.

a) $\dfrac{9}{8}$

b) $\dfrac{17}{5}$

c) $\dfrac{6}{4}$

d) $\dfrac{19}{10}$

e) $\dfrac{18}{16}$

10. Change these mixed numbers into improper fractions.

a) $1\dfrac{2}{3}$

b) $2\dfrac{2}{3}$

c) $1\dfrac{3}{10}$

d) $3\dfrac{1}{5}$

e) $4\dfrac{5}{6}$

11. Add together the following fractions. Always write the answers in the lowest terms.

a) $\dfrac{1}{6} + \dfrac{1}{6} =$

b) $\dfrac{1}{4} + \dfrac{3}{8} =$

c) $\dfrac{2}{3} + \dfrac{5}{6} =$

d) $\dfrac{9}{21} + \dfrac{4}{7} =$

e) $1\dfrac{3}{5} + \dfrac{1}{10} =$

f) $1\dfrac{1}{4} + \dfrac{11}{12} =$

g) $1\dfrac{2}{3} + \dfrac{2}{3} =$

h) $1\dfrac{5}{6} + 3\dfrac{2}{5} =$

12. Subtract the following fractions. Always write the answers in the lowest terms.

a) $\dfrac{3}{4} - \dfrac{1}{4} =$

b) $\dfrac{3}{5} - \dfrac{2}{5} =$

c) $\dfrac{5}{6} - \dfrac{2}{3} =$

d) $\dfrac{5}{7} - \dfrac{1}{14} =$

e) $2\dfrac{1}{2} - 1\dfrac{1}{4} =$

f) $2\dfrac{2}{5} - 2\dfrac{1}{5} =$

g) $3\dfrac{3}{4} - 2\dfrac{1}{2} =$

h) $5\dfrac{2}{3} - 3\dfrac{1}{6} =$

13. Multiply these fractions. Write the answers in the lowest terms.

a) $2 \times \dfrac{2}{5} =$

b) $4 \times \dfrac{1}{7} =$

c) $4 \times \dfrac{2}{3} =$

d) $\dfrac{3}{4} \times \dfrac{2}{3} =$

e) $\dfrac{5}{9} \times \dfrac{3}{5} =$

f) $2\dfrac{1}{3} \times 3\dfrac{1}{2} =$

g) $2\dfrac{1}{5} \times 2\dfrac{1}{2} =$

h) $2\dfrac{1}{2} \times 2\dfrac{1}{2} =$

14. Divide these fractions. Write the answers in the lowest terms.

a) $3 \div \dfrac{1}{2} =$

b) $4 \div \dfrac{1}{3} =$

c) $\dfrac{1}{3} \div \dfrac{2}{3} =$

d) $\dfrac{1}{5} \div \dfrac{3}{10} =$

e) $\dfrac{4}{5} \div \dfrac{2}{3} =$

f) $1\dfrac{3}{4} \div 1\dfrac{2}{3} =$

g) $2\dfrac{1}{5} \div 1\dfrac{1}{2} =$

15. What is $\dfrac{1}{4}$ of 36?

16. What is $\dfrac{2}{3}$ of 27?

17. Copy this number sentence, and fill in the missing number.

$\dfrac{1}{4}$ of _____ = 15

18. A class has 30 pupils. $\frac{2}{5}$ of the class are girls. How many pupils are boys and how many are girls?

19. A fisherman caught 48 fish. The fisherman put eight fish back into the sea because they were very small. What fraction of the fish did he put back into the sea? What fraction did he keep?

20. Rosa went to the shops. She spent $\frac{1}{3}$ of her money at the butcher's and $\frac{1}{2}$ of her money at the book shop. What fraction of her money did she spend altogether? What fraction of her money did she have left?

21. Isabelle visited her cousin's house. On the way to her cousin's house she saw eight cows, five goats, four donkeys, and five dogs. What fraction of the animals she saw were cows?

22. Rachel and David share 40 biscuits between them in the ratio 2:3. How many biscuits do they each get?

7 Decimals

What are Decimals?

So far we have used fractions to describe parts of a whole number. Decimals are another way of describing parts of a whole number. Some historians believe that decimals were first used many centuries ago by an Arab mathematician called al-Uglidisi. Some other historians think that decimals were first used in India.

In Chapter 1 *Numbers and Place Value*, we learnt that the position of each digit in a number tells us how much that digit is worth. For example, the number 246 is made up of 2 hundreds, 4 tens and 6 units. We can write this as:

h	t	u
2	4	6

If we move across the number from right to left, then each position in the number has a value ten times bigger than the position before. A ten is ten times bigger than a unit, and a hundred is tens time bigger than a ten.

With decimals we can use the same idea of place value for fractions of whole numbers.

Hundreds	Tens	Units	Decimal point	Tenths	Hundredths	Thousandths
h	t	u	.	$\frac{1}{10}$	$\frac{1}{100}$	$\frac{1}{1000}$

The hundreds, tens and units are whole numbers. The tenths and hundredths and thousandths are fractions of a whole number. The decimal point is written as a dot, like a full-stop. The decimal point shows us where the whole number ends and the fraction part of the number begins. An example of a decimal number is 2.6, which we say as "two point six". It means "2 units and 6 tenths".

In some countries the decimal point is written as a comma. For example, "two point six" is written as 2,6. But in most countries people use a point. If we look at the number 246.431 we can see that it is made up of 2 hundreds, 4 tens, 6 units, 4 tenths, 3 hundredths and 1 thousandth, as shown in the table on page 77.

Hundreds	Tens	Units	Decimal point	Tenths	Hundredths	Thousandths
h	t	u	.	$\frac{1}{10}$	$\frac{1}{100}$	$\frac{1}{1000}$
2	4	6	.	4	3	1
This digit has the value of 2 hundreds	This digit has the value of 4 tens	This digit has the value of 6 units	.	This digit has the value of 4 tenths	This digit has the value of 3 hundredths	This digit has the value of 1 thousandth
200	40	6	.	$\frac{4}{10}$	$\frac{3}{100}$	$\frac{1}{1000}$

The number 246.431 can be split into:

$$200 + 40 + 6 + \frac{4}{10} + \frac{3}{100} + \frac{1}{1000} = 246.431$$

Changing Fractions into Decimals

We can turn a fraction into a decimal by remembering that a fraction like $\frac{4}{10}$ is the same as $4 \div 10$:

$$\frac{4}{10} = 4 \div 10 = 10\overline{)4.40}^{\,0.\,4}$$

10 doesn't divide into 4 so we write 0 in the answer line above the 4 and put in a decimal point after the 4 *and* above it in the answer line. Then we put a zero to give us 4.0.

Now we add the 4 units to the tenths to give us 40 tenths. This is because 4 units is the same as 40 tenths. We show this by writing a small 4 in front of the 0 in the tenths place to give us 40. Now we divide the 10 into 40. This gives us 4. So we write 4 in the answer line above the 0. The complete answer is 0.4:

$$\frac{4}{10} = 0.4$$

Here is an example using hundredths. Let's change $\frac{3}{100}$ into a decimal:

$$\frac{3}{100} = 3 \div 100 = 100\overline{)3.30300}^{\,0.\,0\,3}$$

100 doesn't divide into 3 so we write 0 in the answer line above the 3 and put in a decimal point after the 3 *and* above it in the answer line. Then we put a zero to give us 3.0.

Now we add the 3 units to the tenths to give us 30 tenths. But 100 still doesn't divide into 30 so we write another 0 in the answer line above the 30 tenths. Then we put another 0 to give us 3.00 and we add the 30 tenths to the hundredths to give us 300 hundredths. Now we divide the 100 into 300. This gives us 3. So we write 3 in the answer line above the hundredths. The answer is 0.03:

$$\frac{3}{100} = 0.03$$

Tenths, hundredths and thousandths all follow this pattern. For example:

$\frac{1}{10} = 0.1$ $\quad$ $\frac{2}{10} = 0.2$ $\quad$ $\frac{3}{10} = 0.3$ $\quad$ $\frac{4}{10} = 0.4$ $\quad$ and so on

$\frac{1}{100} = 0.01$ $\quad$ $\frac{2}{100} = 0.02$ $\quad$ $\frac{3}{100} = 0.03$ $\quad$ $\frac{4}{100} = 0.04$ $\quad$ and so on

$\frac{1}{1000} = 0.001$ $\quad$ $\frac{2}{1000} = 0.002$ $\quad$ $\frac{3}{1000} = 0.003$ $\quad$ $\frac{4}{1000} = 0.004$ $\quad$ and so on

Remember the number 246.431, which we split into:

$$200 + 40 + 6 + \frac{4}{10} + \frac{3}{100} + \frac{1}{1000} = 246.431$$

Now we can see that this number can also be written as:

$$200 + 40 + 6 + 0.4 + 0.03 + 0.001 = 246.431$$

Exercise 1

Write these numbers showing the value for each digit. For example:

$$18.26 = 10 + 8 + 0.2 + 0.06$$

a) 23.95

b) 57.22

c) 8.624

d) 146.83

e) 47.213

f) 245.472

Here are two more examples of changing fractions into decimals. First, let's change $\frac{1}{2}$ into a decimal:

$$\frac{1}{2} = 1 \div 2 = 2\overline{)1.10} \;\; 0.5$$

2 doesn't divide into 1 so we write 0 in the answer line above the 1 and put in a decimal point after the 1 *and* above it in the answer line. Then we put a zero to give us 1.0.

Next we add the 1 unit to the tenths to give us 10 tenths. Now we divide the 2 into 10. This gives us 5. So we write 5 in the answer line above the 0. The answer is 0.5:

$$\frac{1}{2} = 0.5$$

Now let's change $\frac{3}{4}$ into a decimal:

$$\frac{3}{4} = 3 \div 4 = 4\overline{)3.3_020}\ \ ^{0.\ 7\ 5}$$

4 doesn't divide into 3 so we add the 3 units to the tenths to give us 30 tenths. Then 4 divides into 30 seven times with 2 remaining. We then add the 2 remaining tenths to the hundredths to give 20 hundredths. Then 4 divides into 20 five times. So the answer is 0.75:

$$\frac{3}{4} = 0.75$$

Exercise 2

Convert these fractions into decimals.

a) $\dfrac{3}{10}$

b) $\dfrac{7}{10}$

c) $\dfrac{7}{100}$

d) $\dfrac{1}{4}$

e) $\dfrac{1}{5}$

f) $\dfrac{1}{8}$

g) $\dfrac{2}{5}$

h) $\dfrac{3}{1000}$

Putting Decimals in Order

We can show decimals on a number line in exactly the same way as whole numbers. Here is part of the number line between 0 and 1:

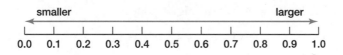

smaller ← → larger

0.0 0.1 0.2 0.3 0.4 0.5 0.6 0.7 0.8 0.9 1.0

We can use the number line to help us put decimals in order. Remember that the numbers on a number line get bigger as we move to the right and smaller as we move to the left. So we can see that 0.4 is bigger than 0.2.

If we have a decimal number such as 0.35 we can show it on the number line half way between 0.3 and 0.4.

Exercise 3

Which number is the biggest?

a) 0.5 or 0.1   b) 0.9 or 0.3

c) 6.2 or 6.4   d) 13.5 or 13.8

e) 4.5 or 4.55   f) 126.65 or 126.56

Adding Decimals

Decimals are added in the same way as whole numbers. When we use place value columns we must make sure that we keep the decimal points one above the other.

For example, if we want to work out 6.4 + 3.2 we can add in columns in the normal way, like this:

```
    6 . 4
 +  3 . 2
    9 . 6
```

If we want to work out 18.35 + 12.2 we can put a 0 at the end of 12.2 so that we have neat columns:

```
   1 8 . 3 5
 + 1 2 . 2 0
         1
   3 0 . 5 5
```

With decimals, putting an extra 0 on the end makes no difference to the number, because it doesn't change the place values of the digits. The decimal point keeps the place values as they are.

Exercise 4

Work out the following. Remember to write them out carefully so that the decimal point is always in the right place.

a) 6.2 + 2.4 =   b) 5.32 + 4.25 =

c) 14.36 + 12.13 =   d) 6.27 + 2.35 =

Exercise 4 *(continued)*

e) $28.97 + 4.25 =$

f) $15.88 + 12.46 =$

g) $98.45 + 14.06 =$

h) $36.35 + 14.4 =$

i) $224.26 + 118.86 =$

j) $245.76 + 182.8 =$

Subtracting Decimals

Decimals are subtracted in the same way as whole numbers. Remember to always keep the decimal points above one another when using columns.

For example, $8.76 - 2.45$ can be worked out like this:

$$
\begin{array}{r}
8\,.\,7\,6 \\
-\;2\,.\,4\,5 \\
\hline
6\,.\,3\,1
\end{array}
$$

We can put a 0 at the end of decimals to give us neat columns and allow us to subtract. For example, $42.5 - 4.35$ is written like this:

$$
\begin{array}{r}
{}^{3}\!\!\not4\;{}^{1}2\,.\,{}^{4}\!\not5\;{}^{1}0 \\
-\qquad 4\,.\,3\,5 \\
\hline
3\;8\,.\,1\,5
\end{array}
$$

So $42.5 - 4.35 = 38.15$.

Exercise 5

Work out the following. Remember to write them out carefully so that the decimal point is in the right place.

a) $5.4 - 3.2 =$

b) $9.8 - 5.3 =$

c) $12.6 - 3.7 =$

d) $33.2 - 16.4 =$

e) $8.54 - 3.21 =$

f) $18.86 - 12.67 =$

g) $26.24 - 8.35 =$

h) $24.6 - 2.42 =$

i) $34.1 - 3.29 =$

j) $132.26 - 44.8 =$

Multiplying Decimals

Multiplying Decimals by 10, 100 and 1000

If we want to multiply a decimal by 10 we do it in exactly the same way as multiplying a whole number by 10. All the digits move one place to the left. For example:

$5.36 \times 10 = 53.6$

With decimals there is no need to put a 0 on the end if we don't need it, so the answer is 53.6.

Here are some other examples:

$8.63 \times 10 = 86.3$
$1.47 \times 10 = 14.7$
$0.27 \times 10 = 2.7$

If we want to multiply a decimal by 100 it is the same as multiplying a whole number by 100. All the digits move two places to the left. For example:

$5.36 \times 100 = 536.0 = 536$

Here are some other examples:

$8.63 \times 100 = 863$
$1.475 \times 100 = 147.5$
$0.273 \times 100 = 27.3$

If we want to multiply a decimal by 1000 all the digits move three places to the left. For example:

$5.36 \times 1000 = 5360.0 = 5360$
$1.475 \times 1000 = 1475$
$0.273 \times 1000 = 273$

Exercise 6

Try these questions.

a) $4.38 \times 10 =$

b) $187.2 \times 10 =$

c) $9.14 \times 10 =$

d) $0.83 \times 10 =$

e) $0.056 \times 10 =$

f) $43.5 \times 100 =$

g) $154.63 \times 100 =$

h) $0.0034 \times 100 =$

i) $8.365 \times 1000 =$

j) $0.0034 \times 1000 =$

Multiplying Decimals by a Whole Number

If we want to multiply a decimal by a different whole number, we can use the same methods that we used to multiply two whole numbers. For example, we could set out 2.14×4 as a short multiplication. We must make sure we keep the decimal points in a vertical line, on top of each other. The number of decimal places in the answer will be the same as the number of decimal places in the question.

$$
\begin{array}{r}
2 \,.\, 1\; 4 \\
\times \quad 4 \\
\hline
\end{array}
$$

$0.04 \times 4 =$	$0 \,.\, 1\; 6$
$0.10 \times 4 =$	$0 \,.\, 4\; 0$
$2.00 \times 4 =$	$8 \,.\, 0\; 0$
then add the answers	$8 \,.\, 5\; 6$

So $2.14 \times 4 = 8.56$.

Here is another example:

$5.92 \times 3 =$

$$
\begin{array}{r}
5 \,.\, 9\; 2 \\
\times \quad 3 \\
\hline
\end{array}
$$

$0.02 \times 3 =$	$0 \,.\, 0\; 6$
$0.90 \times 3 =$	$2 \,.\, 7\; 0$
$5.00 \times 3 =$	$1\, 5 \,.\, 0\; 0$
then add the answers	$1\, 7 \,.\, 7\; 6$

So $5.92 \times 3 = 17.76$.

Here is an example of a decimal multiplied by a bigger whole number:

$2.25 \times 14 =$

$$
\begin{array}{r}
2 \,.\, 2\; 5 \\
\times 1\, 4 \\
\hline
\end{array}
$$

$2.25 \times 4 =$	$9 \,.\, 0\; 0$
$2.25 \times 10 =$	$2\, 2 \,.\, 5\; 0$
then add the answers	$3\, 1 \,.\, 5\; 0$

So $2.25 \times 14 = 31.50$.

Multiplying Decimals by Another Decimal

If we want to multiply a decimal number by another decimal number we multiply in the usual way. The number of decimal places in the answer will be the same as the total number of decimal places in the question. For example:

$$
\begin{array}{r}
2 \,.\, 3 \qquad \text{one decimal place} \\
\times \quad 1 \,.\, 5 \qquad \text{one decimal place} \\
\hline
1\; 1\; 5 \qquad\qquad\qquad \\
2\; 3\; 0 \qquad\qquad\qquad \\
\hline
3 \,.\, 4\; 5 \qquad \text{two decimal places} \\
\end{array}
$$

Exercise 7

Here are some multiplications for you to try.

a) $1.21 \times 4 =$

b) $6.25 \times 3 =$

c) $4.54 \times 6 =$

d) $14.2 \times 5 =$

e) $3.34 \times 12 =$

f) $2.4 \times 30 =$

g) $12.2 \times 12 =$

h) $14.64 \times 14 =$

i) $3.4 \times 1.2 =$

j) $0.5 \times 0.5 =$

k) $3.6 \times 1.5 =$

l) $12.4 \times 0.4 =$

Dividing Decimals

Dividing Decimals by 10, 100 and 1000

If we want to divide a decimal number by 10, 100 or 1000 we use exactly the same method as dividing a whole number by 10, 100 or 1000. All the digits move one place to the right when dividing by 10, two places to the right when dividing by 100, and three places to the right when dividing by 1000. For example:

$23.7 \div 10 = 2.37$

$142.67 \div 10 = 14.267$

$9 \div 10 = 0.9$

$2342 \div 100 = 23.42$

$86.5 \div 100 = 0.865$

$152 \div 100 = 1.52$

$2342 \div 1000 = 2.342$

$8564 \div 1000 = 8.564$

$157 \div 1000 = 0.157$

Exercise 8

Work out the following.

a) $32.8 \div 10 =$

b) $1786 \div 10 =$

c) $1.356 \div 10 =$

d) $0.354 \div 10 =$

e) $256 \div 100 =$

f) $1385 \div 100 =$

g) $23.7 \div 100 =$

h) $8.635 \div 100 =$

i) $4361 \div 1000 =$

j) $1762 \div 1000 =$

Dividing Decimals by a Whole Number

If we want to divide a decimal by a different whole number we can use short or long division to help us.

For example, $27.6 \div 6$ can be worked out using short division:

$$\begin{array}{r} 4.\;6 \\ 6\overline{)2\;7.{}_36} \end{array}$$

6 divides into 27 four times with 3 remaining. We write 4 on the answer line above, and a small 3 in front of the 6 to give us 36 tenths. This is because we have added the 3 units remaining to the tenths. 3 units is the same as 30 tenths so now we have 36 tenths altogether.

Now 6 divides into 36 exactly 6 times, so we write 6 on the answer line giving us an answer of 4.6. Remember to keep the decimal points one above the other.

Exercise 9

Work out the answers to these divisions.

a) $18.6 \div 3 =$

b) $12.8 \div 4 =$

c) $5.6 \div 2 =$

d) $99.9 \div 9 =$

e) $17.4 \div 3 =$

f) $149.1 \div 7 =$

g) $131.2 \div 8 =$

h) $39.5 \div 5 =$

i) $112.2 \div 6 =$

j) $35.2 \div 11 =$

Dividing Decimals by Another Decimal

In Chapter 4 *Division* we discovered that if we multiply both numbers in a division by the same number we get the same answer. So $16 \div 4$ is the same as $160 \div 40$ (both numbers multiplied by 10). In both cases the answer is 4.

So, if we want to divide a decimal by another decimal, we can make the division easier by multiplying both sides by 10 or 100 until we are dividing by a whole number. For example, $2.5 \div 0.5$ is the same as $25 \div 5$ (both numbers multiplied by 10). So $2.5 \div 0.5 = 5$.

Here is another example: $12.4 \div 0.8$ is the same as $124 \div 8$ (both numbers have been multiplied by 10). Now we are dividing by a whole number, which is much easier to work out:

$$\begin{array}{r} 1\;5.\;5 \\ 8\overline{)1\;2{}_44.{}_40} \end{array}$$

So $12.4 \div 0.8 = 15.5$.

Exercise 10

Work out the answers to these divisions.

a) $3.5 \div 0.5 =$

b) $7.8 \div 1.2 =$

c) $22.1 \div 1.7 =$

d) $1.21 \div 1.1 =$

e) $8.4 \div 0.8 =$

f) $8.32 \div 2.6 =$

g) $1.44 \div 0.02 =$

h) $1.26 \div 0.35 =$

Showing Remainders as Decimals

We saw in Chapter 4 *Division* that sometimes when we divide a whole number by another whole number we are left with a remainder. For example:

$15 \div 6 = 2 \text{ r}3$

In the examples above we have been showing remainders as decimals, so we can now see that if we have a question like $15 \div 6$ we could write the answer as 2 r3 or we could show the answer as a decimal like this:

$$\begin{array}{r} 2.\,5 \\ 6\overline{)1\,5.{}_30} \end{array}$$

So $15 \div 6 = 2 \text{ r}3$ or 2.5.

Here is another example:

$26 \div 5 = 5 \text{ r}1$

Or we can write it as a decimal:

$$\begin{array}{r} 5.\,2 \\ 5\overline{)2\,6.{}_10} \end{array}$$

So $26 \div 5 = 5 \text{ r}1$ or 5.2.

Sometimes the remainder will be more than 1 decimal place, like this example:

$34 \div 8 =$

$$\begin{array}{r} 4.\,2\,5 \\ 8\overline{)3\,4.{}_20{}_40} \end{array}$$

8 divides into 34 four times with 2 remaining. We write 4 in the answer line above the 4 and put a decimal point and 0 after 34. Now we write 2 in front of the 0 to show that we have added the 2 remaining units to the tenths to give 20 tenths. Now 8 divides into 20 twice with 4 remaining. So we write 2 in the answer line

above the 0 after the decimal point. Then we put another 0 and write 4 in front of it to show that we have added the 4 tenths remaining to the hundredths to give 40 hundredths. Now 8 divides into 40 exactly 5 times, so we write 5 above the second 0 to give us an answer of 4.25:

$$34 \div 8 = 4.25$$

Exercise 11

Work out the following divisions, giving the remainders as decimals.

a) $9 \div 2 =$ b) $10 \div 4 =$

c) $21 \div 6 =$ d) $32 \div 5 =$

e) $18 \div 8 =$ f) $54 \div 5 =$

g) $69 \div 6 =$ h) $46 \div 8 =$

Rounding

Rounding to 2 or 3 Decimal Places

Sometimes we find that the remainder works out to be more than 2 decimal places. For example if we work out $30 \div 9$ we find that the answer is 3.3333... with the decimal places running on forever! Try it for yourself. Normally we decide that it is sensible to stop the remainder after 2 or 3 decimal places. So for $30 \div 9$ we would write 3.33 or 3.333.

Stopping the decimals after a number of decimal places is called **rounding**. If we want to round a number to 2 decimal places we look at the third decimal place to see whether the digit there is less than 5. If it is less than 5 then we just stop the number at the second decimal place.

For example:

    2.632 rounded to 2 decimal places is 2.63
    17.841 rounded to 2 decimal places is 17.84
    6.284 rounded to 2 decimal places is 6.28

In each case the digit in the third decimal place is less than 5.

If the digit in the third decimal place is 5 or greater, then we round up the second decimal place to the next digit.

For example, if we were to look at 3.258 on a number line, we would see that it is closer to 3.26 than it is to 3.25. So 3.258 rounded to 2 decimal places is 3.26.

Here are some more examples:

    1.239 rounded to 2 decimal places is 1.24
    26.527 rounded to 2 decimal places is 26.53
    5.235 rounded to 2 decimal places is 5.24

If we want to round to 3 decimal places we look at the digit in the fourth decimal place to see if it is less than 5, or not less than 5.

For example:

    1.2443 rounded to 3 decimal places is 1.244
    7.8247 rounded to 3 decimal places is 7.825
    14.8565 rounded to 3 decimal places is 14.857

Exercise 12

Round these decimals to 2 decimal places.

a) 4.561   b) 8.456

c) 17.915   d) 12.382

e) 5.685   f) 0.463

Round these decimals to 3 decimal places.

g) 9.0241   h) 8.4937

i) 1.3408   j) 21.6221

k) 13.8015   l) 0.0007

Remember!

◆   We can use place value to show fractions as decimals.
◆   We can show decimals on a number line in the same way as whole numbers.
◆   We can add, subtract, multiply and divide decimals in the same way as whole numbers but we must remember to keep the decimal point in the right place.
◆   We can show remainders as decimals.
◆   If we need to, we can add one or more 0s to the end of a decimal.
◆   When we multiply a decimal by a decimal, the number of decimal places in the answer will be the same as the total number of decimal places in the question.
◆   We can make it easier to divide a decimal by another decimal by multiplying both numbers by 10 or 100 until we are dividing by a whole number.
◆   We can write decimals to a certain number of decimal places by rounding up or down.

Revision Test   on Decimals

Now that you have worked your way through the chapter, try this revision test. The answers are in the answer book.

1. Write these fractions as decimals.

a) $\dfrac{2}{10}$ b) $\dfrac{4}{10}$

c) $\dfrac{7}{100}$ d) $\dfrac{45}{100}$

e) $\dfrac{4}{5}$ f) $\dfrac{3}{8}$

g) $\dfrac{1}{20}$

2. Put these decimals in order of size from the smallest to the biggest:

1.65, 2.34, 0.85, 1.56, 0.09, 5.21, 0.83

3. a) $2.32 + 1.63 =$

b) $18.42 + 26.39 =$

c) $4.54 + 0.27 + 0.04 =$

d) $47.52 + 2.031 =$

4. a) $8.6 - 3.2 =$

b) $17.48 - 8.31 =$

c) $9.6 - 0.34 =$

d) $132.58 - 12.59 =$

5. a) $12.43 \times 10 =$

b) $0.078 \times 10 =$

c) $1.746 \times 100 =$

d) $13.63 \times 100 =$

6. a) $8.12 \times 4 =$

b) $15.23 \times 6 =$

c) $23.06 \times 12 =$

d) $127.73 \times 5 =$

7. a) $16.2 \times 2.4 =$

b) $0.6 \times 0.7 =$

c) $2.5 \times 3.7 =$

d) $8.4 \times 3.6 =$

$\rightarrow$

Revision Test on Decimals *(continued)*

8. Work out the answers to these division problems, giving the remainders as decimals:

   a) $14 \div 4 =$

   b) $39 \div 6 =$

   c) $86 \div 8 =$

   d) $28 \div 5 =$

9. a) $23.6 \div 10 =$

   b) $167.89 \div 10 =$

   c) $0.65 \div 10 =$

   d) $72 \div 10 =$

10. a) $867.2 \div 100 =$

    b) $67.5 \div 100 =$

    c) $0.4 \div 100 =$

    d) $11 \div 100 =$

11. a) $28.8 \div 2 =$

    b) $66.24 \div 3 =$

    c) $0.8 \div 5 =$

    d) $78.8 \div 4 =$

12. a) $3.6 \div 0.6 =$

    b) $5.46 \div 1.3 =$

    c) $0.783 \div 0.09 =$

    d) $50.6 \div 2.2 =$

13. Round these decimals to 2 decimal places:

    a) 8.091

    b) 6.123

    c) 0.905

    d) 26.347

14. Round these decimals to 3 decimal places:

    a) 0.0131

    b) 2.3465

    c) 11.6666

    d) 11.3333

8 Percentages

What is a Percentage?

A percentage is a fraction where the bottom number (the denominator) is always 100. We use the symbol % to show a percentage.

For example, $\frac{25}{100}$ is written as 25%. We say this as "25 percent".

So if a fraction has a bottom number of 100, then the top number is the percentage. Here are three more examples:

$$\frac{70}{100} = 70\% \qquad\qquad \frac{10}{100} = 10\% \qquad\qquad \frac{55}{100} = 55\%$$

If we want to show a fraction as a percentage when the bottom number isn't 100, then we multiply the top and bottom numbers by the same number until the bottom number is 100.

For example, to show $\frac{25}{50}$ as a percentage, we multiply the top and bottom numbers by 2 so that the bottom number becomes 100:

$$\frac{25}{50} = \frac{50}{100} = 50\%$$

Here are four more examples:

$$\frac{35}{50} = \frac{70}{100} = 70\% \qquad \text{top and bottom multiplied by 2}$$

$$\frac{12}{25} = \frac{48}{100} = 48\% \qquad \text{top and bottom multiplied by 4}$$

$$\frac{6}{10} = \frac{60}{100} = 60\% \qquad \text{top and bottom multiplied by 10}$$

$$\frac{1}{2} = \frac{50}{100} = 50\% \qquad \text{top and bottom multiplied by 50}$$

Exercise 1

Write these fractions as percentages.

a) $\dfrac{85}{100}$ b) $\dfrac{17}{50}$

c) $\dfrac{7}{20}$ d) $\dfrac{21}{25}$

e) $\dfrac{3}{10}$ f) $\dfrac{2}{5}$

g) $\dfrac{9}{10}$ h) $\dfrac{3}{4}$

Comparing Fractions, Decimals and Percentages

We know from Chapter 7 *Decimals* that we can turn fractions into decimals, and we have just seen that we can also turn fractions into percentages. So now we can compare fractions, decimals and percentages.

For example, we know that $\frac{1}{4} = 0.25$, because:

$$\begin{array}{r} 0.2\ 5 \\ 4\overline{)1.0_20} \end{array}$$

We also know that:

$$\frac{1}{4} = 25\% \qquad \text{top and bottom multiplied by 25 gives } \frac{25}{100}$$

So $\frac{1}{4}$, 0.25 and 25% are all the same. They are **equivalent**.

$$\frac{1}{4} = 0.25 = 25\%$$

Here are some more examples:

$$\frac{1}{4} = 0.25 = 25\% \qquad\qquad \frac{1}{2} = 0.5 = 50\%$$

$$\frac{3}{4} = 0.75 = 75\% \qquad\qquad \frac{4}{5} = 0.8 = 80\%$$

$$\frac{1}{10} = 0.1 = 10\% \qquad\qquad \frac{2}{10} = 0.2 = 20\%$$

$$\frac{3}{10} = 0.3 = 30\% \qquad\qquad \frac{5}{10} = 0.5 = 50\%$$

$$\frac{1}{100} = 0.01 = 1\% \qquad\qquad \frac{2}{100} = 0.02 = 2\%$$

$$\frac{75}{100} = 0.75 = 75\% \qquad\qquad 1 = \frac{1}{1} = 1.00 = 100\%$$

If we look at all the examples above we can see that in every case the percentage is the same as the decimal multiplied by 100. So if we start with a decimal we can turn it into a percentage by multiplying it by 100. For example:

$0.2 \times 100 = 20$, so 0.2 is the same as 20%
$0.45 \times 100 = 45$, so 0.45 is the same as 45%
$0.63 \times 100 = 63$, so 0.63 is the same as 63%

Multiplication and division are opposites, so if we start with a percentage we can turn it into a decimal by dividing by 100.

For example, to change 20% into a decimal, divide 20 by 100:

$20 \div 100 = 0.2$, so 20% is the same as 0.2

If we have the whole of anything it is the same as 100%. For example:

$$1 = \frac{1}{1} = 1.00 = 100\%$$

If we have a cake divided into 8 equal pieces and we have all the pieces, we have:

$$\frac{8}{8} = \frac{1}{1} = 1.00 = 100\%$$

So we have 100% of the cake. If we have 100% of something, it means we have all of it.

Exercise 2

Copy the table and fill in the gaps.

Fraction	Decimal	Percentage
$\frac{7}{10}$		70%
	0.3	30%
$\frac{1}{5}$	0.2	
		90%
$\frac{3}{100}$		3%
	0.75	
$\frac{41}{100}$		
		17%
$\frac{1}{4}$		
	0.23	

Finding a Percentage of an Amount

It is often useful to find a percentage of an amount. Sometimes we can work out a percentage of an amount quickly in our heads by remembering a few simple rules.

♦ 10% is the same as $\frac{1}{10}$ so to find 10% of a number we divide by 10.
♦ 5% is half of 10% so to find 5% we can first find 10% and then halve it.
♦ 1% is the same as $\frac{1}{100}$ so to find 1% of a number we divide by 100.

Remember also that 25% is $\frac{1}{4}$, 50% is $\frac{1}{2}$ and 75% is $\frac{3}{4}$.

For example, if we want to work out 10% of 70 we divide 70 by 10. The answer is 7.

If we are asked to work out 30% of 70, we can divide 70 by 10 to give us 10% of 70, and then multiply the answer by 3 to give us 30% of 70:

$$10\% \text{ of } 70 \text{ is } 70 \div 10 = 7$$
$$30\% = 7 \times 3$$
$$= 21$$

So 30% of 70 is 21.

Here is another example:

"What is 42% of 60?"

First we can work out 10% of 60 = 6.

We can also work out that 1% of 60 = 0.6.

Now we can work out 40% and 2%:

$$40\% \text{ of } 60 = 4 \times 10\% \text{ of } 60 = 4 \times 6 = 24$$
$$2\% \text{ of } 60 = 2 \times 1\% \text{ of } 60 = 2 \times 0.6 = 1.2$$

Now we add them together to give us 42%:

$$42\% \text{ of } 60 = 40\% + 2\% = 24 + 1.2 = 25.2$$

So 42% of 60 is 25.2.

If we are asked to work out 25% or 75% of a number we can think of it as $\frac{1}{4}$ or $\frac{3}{4}$:

"What is 75% of 16?"

$75\% = \frac{3}{4}$ so 75% of 16 is the same as $\frac{3}{4}$ of 16:

$$\frac{3}{4} \times 16 = 12$$

So 75% of 16 is 12.

Often questions like this are written as word problems, like this example:

"35% of class 6 like swimming. There are 40 children in the class. How many children like swimming?"

To answer this problem we need to work out 35% of 40:

10% of 40 is 4, so 30% must be $3 \times 4 = 12$
10% of 40 is 4, so 5% must be 2

To get 35% we need to add them together:

$$35\% \text{ of } 40 = 30\% + 5\% = 12 + 2 = 14$$

So 14 children like swimming.

Exercise 3

Find the following amounts.

a) 20% of 40

b) 15% of 80

c) 75% of 20

d) 12% of 50

e) 55% of 16

f) 71% of 90

g) 23% of 80

h) 35% of 10

i) A school has 420 students. 55% of them are girls and 45% are boys. How many girls are there? How many boys are there?

Sometimes it is too difficult to work out a percentage of an amount in our heads, so instead we have to write out a calculation.

For example, if we want to work out 27% of 130 we can work it out by remembering that 27% is $\frac{27}{100}$ so 27% of 130 is the same as $130 \times \frac{27}{100}$:

$$130 \times \frac{27}{100} = \frac{130}{1} \times \frac{27}{100}$$

$$= \frac{130 \times 27}{1 \times 100}$$

$$= \frac{130 \times 27}{100}$$

You could use any method that works for you to work out 130 × 27, such as the grid method or the vertical method. If you're not sure look again at Chapter 3 Multiplication.

$$= \frac{3510}{100}$$

$$= 35.1$$

So 27% of 130 is 35.1.

Here is another example:

"What is 13% of 730?"

$$730 \times \frac{13}{100} = \frac{730}{1} \times \frac{13}{100}$$

$$= \frac{730 \times 13}{1 \times 100}$$

$$= \frac{730 \times 13}{100}$$

$$= \frac{9490}{100}$$

$$= 94.9$$

So 13% of 730 is 94.9.

Sometimes we need to work out a percentage of a number that is greater than 100%. So we need to know what a percentage greater than 100% means. We know that:

$$\frac{1}{1} = \frac{100}{100} = 100\%$$

This means that:

$$\frac{2}{1} = \frac{200}{100} = 200\%$$

So if we work out 200% of a number we are **doubling** it, which means making it twice as much.

Similarly:

$$\frac{3}{1} = \frac{300}{100} = 300\%$$

So if we work out 300% of a number we are **trebling** it, which means making it three times as much.

For example, to work out 300% of 15 we can write it out in the usual way:

$$15 \times \frac{300}{100} = \frac{15}{1} \times \frac{300}{100} = \frac{15 \times 300}{1 \times 100} = \frac{15 \times 300}{100}$$

As we learnt in Chapter 6 *Fractions and Ratios*, when we divide the top and bottom of a fraction by the same number, we get a fraction of the same size. So here we can divide the top and the bottom by 100, which makes the calculation simpler but doesn't change the answer:

$$\frac{15 \times 300}{100} = \frac{15 \times 3}{1} = 45$$

So 300% of 15 is 45.

Exercise 4

Find the following amounts.

a) 11% of 120   b) 22% of 43

c) 15% of 860   d) 19% of 70

e) 200% of 61   f) 400% of 14

Working Out an Amount as a Percentage

Another use for percentages is to tell us a number as a percentage of a whole. For example, if there is a cake divided into 8 equal pieces and we have 6 of them, what percentage of the cake do we have?

We know that we can take the fraction that we have and turn it into a percentage by multiplying the top and bottom numbers by the same number until the bottom number is 100.

But in our example we have 6 pieces of cake out of 8, so our fraction is $\frac{6}{8}$.

It is difficult to work this out as a fraction with a bottom number of 100, so we can use a quicker way.

The quicker way is to take the fraction of the cake we have and multiply the top number by 100 and then divide by the bottom number to give us the percentage:

$$\frac{6}{8} \times 100\% = \frac{3}{4} \times 100\% = \frac{300}{4}\% = 75\%$$

First we changed $\frac{6}{8}$ into its lowest terms, which is $\frac{3}{4}$.

Then we multiplied the top number by 100 to give us $\frac{300}{4}$.

300 divided by 4 gives us 75, so the answer is that we have 75% of the cake.

Here is another example:

"Michael got 18 questions right out of 25 in his maths test. What percentage did he get right?"

We need to change "18 out of 25" into a percentage.

We could multiply the top and bottom of the fraction $\frac{18}{25}$ by 4 to give us the fraction with 100 as the bottom number:

$$\frac{18}{25} = \frac{72}{100} = 72\%$$

Or we could take $\frac{18}{25}$ and multiply the top number by 100 and then divide by the bottom number to give us the percentage:

$$\frac{18}{25} \times 100\% = \frac{18 \times 100}{25}\%$$

We can simplify the fraction by dividing the top and bottom by 25, to give:

$$\frac{18 \times 4}{1}\% = \frac{72}{1}\% = 72\%$$

So Michael got 72% of the test questions right.

Exercise 5

Now try these questions.

a) What is 5 as a percentage of 20?

b) What is 9 as a percentage of 45?

c) Rebecca scored 60 out of 75 in her end-of-year test. What percentage mark did her teacher give her?

d) There are 35 children in a class. 14 of the children say that Science is their favourite subject at school. What percentage say that Science is their favourite subject?

Remember!

- ◆ A percentage is a fraction where the bottom number is 100.
- ◆ To turn a fraction into a percentage we multiply the top and bottom numbers by the same thing until the bottom number is 100. The top number is then the percentage. Or we can multiply the top number of the fraction by 100 and then divide by the bottom number.
- ◆ To turn a decimal into a percentage we multiply it by 100.
- ◆ To turn a percentage into a decimal we divide it by 100.
- ◆ We can compare fractions, decimals and percentages as they are different ways of writing the same thing.
- ◆ To find the percentage of an amount we multiply the amount by the percentage we want. For example, if we want to find 73% of a number we multiply the number by $\frac{73}{100}$.

Revision Test | on Percentages

Now that you have worked your way through the chapter, try this revision test. The answers are in the answer book.

1. Write these fractions as percentages.

a) $\dfrac{8}{10}$

b) $\dfrac{3}{5}$

c) $\dfrac{16}{25}$

d) $\dfrac{37}{50}$

e) $\dfrac{91}{100}$

Revision Test   on Percentages *(continued)*

2. Write these decimals as percentages.

    a) 0.5   b) 0.25

    c) 0.3   d) 0.75

    e) 0.8

3. Find the following amounts.

    a) 10% of 40   b) 15% of 60

    c) 75% of 40   d) 22% of 50

    e) 12% of 80   f) 33% of 120

    g) 200% of 47   h) 500% of 20

4. A school choir has 25 members. 40% of the choir members are boys.
    How many boys are there in the choir? How many girls are there?

5. A farmer has 60 animals altogether. 55% of his animals are cows.
    How many cows does he have?

6. What is 8 as a percentage of 80?

7. What is 12 as a percentage of 60?

8. What is 7 as a percentage of 28?

9. What is 24 as a percentage of 480?

10. In a village by the sea, 36 out of 48 children like fishing. What percentage
    of the children like fishing? What is this percentage written as a fraction?

11. A football team has won ten games, drawn four games and lost six games.
    What percentage of games has the football team won? What percentage has
    it drawn? What percentage has it lost?

12. If 73% of children pass their end-of-year Maths test, what percentage fail?

Measurement

9   Length, Mass and Capacity

What is Measurement?

In Chapters 1 to 8, we saw how we use numbers to count. Counting tells us how many things there are, or how many of something we have. But we can't use numbers on their own to tell us the size of something or how much there is of something.

If we want to know the size of something or how much there is of something, we have to **measure** it.

- If we want to know how long something is we measure its length.
- If we want to know how heavy something is we measure its mass.
- If we want to know how much water we can put inside a bottle we measure the bottle's capacity.

Length, mass and capacity are all measures. We use measures to tell us how much of something there is. When we measure the length of something, or measure the mass of something, or find out the capacity of something, we are taking a **measurement**.

How Do We Measure Length?

We measure length in millimetres, centimetres, metres and kilometres. These are known as **units** of length. To save time when writing we use these letters for each one:

mm for millimetres
cm for centimetres
m for metres
km for kilometres

- 1 millimetre (1 mm) is about the same size as the point of a pen or pencil.
- 1 centimetre (1 cm) is about as wide as a fingernail.
- 1 metre (1 m) is about the length of a long step taken by an adult when walking.

- 1000 millimetres are the same length as 1 metre, so 1000 mm = 1 m.
- 100 centimetres are the same length as 1 metre, so 100 cm = 1 m.
- 1000 metres are the same length as 1 kilometre, so 1000 m = 1 km.

In everyday life we normally use a ruler or a tape measure to measure length. For longer distances we might use a measuring wheel or even the distance measurer in a car.

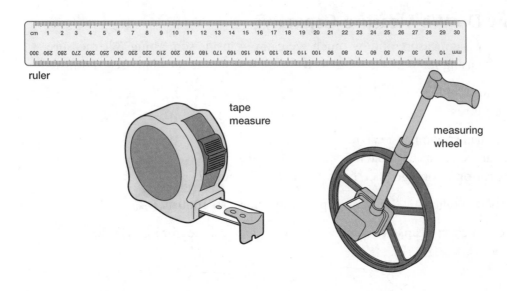

ruler

tape measure

measuring wheel

How Do We Measure Mass?

We measure mass using grams, kilograms and tonnes. These are the units of mass we use most often. People often use the word **weight** when really they mean **mass**. You will learn more about weight and mass in your Science lessons. At the moment all you need to know is that when you weigh something you are finding its mass.

We use these letters for the units of mass:

g for grams
kg for kilograms
t for tonnes

◆ A matchstick has a mass of about 1 gram (1 g).
◆ A newborn baby has a mass of about 3 or 4 kilograms (3 kg or 4 kg).
◆ A small car has a mass of about 1 tonne (1 t).

◆ 1000 grams is the same mass as 1 kilogram, so 1000 g = 1 kg.
◆ 1000 kilograms is the same mass as 1 tonne, so 1000 kg = 1 t.

We can measure mass using a set of scales or a spring balance.

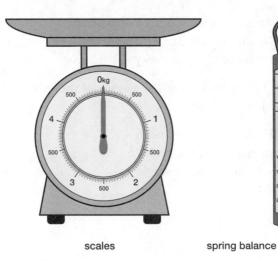

scales

spring balance

How Do We Measure Capacity?

Capacity is a measure of how much we can put inside something. We measure capacity using millilitres and litres. We use these letters for millilitres and litres:

m*l* for millilitres
l for litres

◆ A small spoon has a capacity of about 5 m*l*.
◆ A drink can has a capacity of about 330 m*l*.
◆ A carton of orange juice has a capacity of about 1 *l*.

◆ 1000 millilitres is the same capacity as 1 litre, so 1000 m*l* = 1 *l*.

We can measure capacity using a measuring jug or a measuring cylinder.

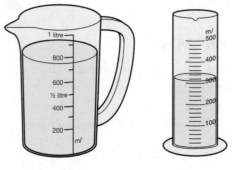

measuring jug   measuring cylinder

The Metric System

The way we measure length, mass and capacity is known as the **metric system**. For each type of measurement there is a main unit.

◆ The main unit for length is the metre.
◆ The main unit for mass is the gram.
◆ The main unit for capacity is the litre.

We can tell by looking at the names of the other units whether they are bigger or smaller than the main unit.

◆ **milli**- means a thousandth
◆ **centi**- means a hundredth
◆ **kilo**- means a thousand

So, for example, a millimetre is $\frac{1}{1000}$ of a metre because "milli" means a thousandth.

A centimetre is $\frac{1}{100}$ of a metre because "centi" means a hundredth.

A kilogram is 1000 grams because "kilo" means a thousand.

Exercise 1

Copy and complete this table.

Name	Letters used	Value in main unit
millimetre		$\frac{1}{1000}$ metre
centimetre	cm	
metre	m	1 metre
kilometre	km	
gram	g	1 gram
	kg	1000 grams
millilitre	m*l*	
litre		1 litre

In the metric system, some of the units for different types of measurements are linked to each other.

For example, 1 millilitre of water has a mass of 1 gram. This means that 1 litre of water has a mass of 1 kilogram. It is very useful to remember this:

1 *l* of water has a mass of 1 kg

Changing Units

Sometimes we want to change a measurement from one unit into another unit. For example, we might want to change a measurement in centimetres into a measurement in metres.

To change a unit to a bigger unit we *divide*, because there will be *fewer* of the bigger unit.

To change a unit to a smaller unit we *multiply*, because there will be *more* of the smaller unit.

Changing Units of Length

If we want to change from one of the units of length into another unit of length, we have to multiply or divide like this:

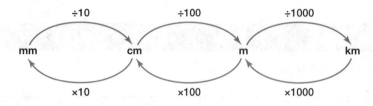

So, for example, if we want to change 50 millimetres into centimetres, we must divide by 10. (Remember from Chapter 4 *Division* that when we divide by 10 the digits move one place to the right.)

$$50 \div 10 = 5$$

So 50 mm = 5 cm.

If we want to change 4 kilometres into metres we must multiply by 1000. (Remember from Chapter 3 *Multiplication* that when we multiply by 1000 the digits move three places to the left.)

$$4 \times 1000 = 4000$$

So 4 km = 4000 m.

Exercise 2

Complete these changes of units. Look at the diagram above to help you remember whether to multiply or divide.

a) 8000 m = _____ km

b) 4 cm = _____ mm

c) 20 mm = _____ cm

d) 300 cm = _____ m

e) 4.5 km = _____ m

f) 620 cm = _____ m

g) 34.6 mm = _____ cm

h) 7.1 m = _____ cm

i) 8.54 cm = _____ mm

j) 1860 m = _____ km

Changing Units of Mass

If we want to change from one of the units of mass into another unit of mass, we have to multiply or divide in the same way we did to change units of length. So to change a unit of mass into a bigger unit of mass we divide, and to change a unit of mass into a smaller unit of mass we multiply.

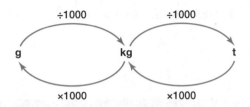

For example, if we want to change 5 kilograms into grams we multiply by 1000:

$$5 \times 1000 = 5000$$

So 5 kg = 5000 g.

If we want to change 526 kilograms into tonnes we divide by 1000. (Remember from Chapter 4 *Division* that when we divide any number by 1000 the digits move three places to the right.)

$$526 \div 1000 = 0.526$$

So 526 kg = 0.526 t.

Exercise 3

Complete these changes of units. Look at the diagram above to help you remember whether to multiply or divide.

a) 10 kg = _____ g

b) 800 g = _____ kg

c) 0.76 t = _____ kg

d) 0.35 kg = _____ g

e) 1.5 kg = _____ g

f) 4 t = _____ kg

g) 2300 g = _____ kg

h) 175 g = _____ kg

i) 150 kg = _____ g

j) 150 kg = _____ t

Changing Units of Capacity

If we want to change millilitres into litres we divide by 1000, and if we want to change litres into millilitres we multiply by 1000.

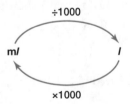

For example, if we want to change 800 millilitres into litres we divide by 1000:

$$800 \div 1000 = 0.8$$

So 800 m*l* = 0.8 *l*.

If we want to change 0.8 litres into millilitres we multiply by 1000:

$0.8 \times 1000 = 800$

So $0.8 l = 800\,\text{m}l$.

Exercise 4

Complete these changes of units.

a) $3000\,\text{m}l = $ _____ l

b) $2l = $ _____ $\text{m}l$

c) $6.5 l = $ _____ $\text{m}l$

d) $1852\,\text{m}l = $ _____ l

e) $0.652 l = $ _____ $\text{m}l$

f) $750\,\text{m}l = $ _____ l

g) $0.006 l = $ _____ $\text{m}l$

Word Problems

Questions about measurement are often given as word problems, like this example:

"Jerome ran four times around a 400-metre running track. How far did he run in metres? How far did he run in kilometres?"

If Jerome ran four times round the 400 m track then the distance he ran is:

$400\,\text{m} \times 4 = 1600\,\text{m}$

To change metres into kilometres we divide by 1000:

$1600 \div 1000 = 1.6$

So Jerome ran 1600 m, which is the same as 1.6 km.

Here is a second example:

"Jane puts 10 g of sugar in her tea. She drinks three cups of tea each day. How many grams of sugar does she put in her tea each day? How much is this in kilograms?"

If Jane puts 10 g of sugar in each cup of tea and she drinks three cups a day then the amount of sugar she uses each day is:

$10\,\text{g} \times 3 = 30\,\text{g}$

To change grams to kilograms we divide by 1000:

$30 \div 1000 = 0.03$

So Jane uses 30 g of sugar each day, which is the same as 0.03 kg.

Let's look at one more example:

"Four litres of water is poured equally into ten cups. How much water is there in each cup in litres? How much is this in millilitres?"

To work this out we need to share 4 litres equally between 10 cups, which means dividing 4 by 10:

$4 \div 10 = 0.4$

So each cup has $0.4\,l$ of water.

To change from litres to millilitres we need to multiply by 1000:

$0.4 \times 1000 = 400$

So each cup has $0.4\,l$ of water, which is the same as $400\,ml$.

Exercise 5

Find the answers to these word problems.

a) Sally wants to buy 1.6 m of fabric to make a dress. When she gets to the shop she finds that the fabric is sold in centimetres. How many centimetres of fabric does she need?

b) John wants to make a cake for his mother. He needs 1.5 kg of sugar to make the cake. Each bag of sugar in the shop has a mass of 500 g. How many bags does he need to buy?

c) A ship sailed for 180 km from one country to another. It then sailed for 5500 m along the coast to the next port. How far did it sail altogether in kilometres?

d) Sunil had 1 l of water. He drank 450 ml of the water. How many millilitres did he have left?

e) Robert had a jug with a mass of 0.5 kg. He poured in some water and then measured the mass again. The jug and water together had a mass of 4.5 kg. How many litres of water did he pour into the jug?

f) A box was loaded with 3 kg of flour. Another 420 g of flour was added to the box. How much flour was then in the box in kilograms?

Remember!

♦ Counting tells us how many. Measuring tells us how much.
♦ Length, mass and capacity are all types of measure.
♦ The main unit of length is the metre (m), the main unit of mass is the gram (g), and the main unit of capacity is the litre (*l*).

$\rightarrow$

Remember! *(continued)*

- ◆ "milli" means a thousandth.
- ◆ "centi" means a hundredth.
- ◆ "kilo" means a thousand.
- ◆ To change a unit to a bigger unit we *divide*, because there will be *fewer* of the bigger unit.
- ◆ To change a unit to a smaller unit we *multiply*, because there will be *more* of the smaller unit.
- ◆ 1 litre of water has a mass of 1 kilogram.

Revision Test   on Length, Mass and Capacity

Now that you have worked your way through the chapter, try this revision test. The answers are in the answer book.

1. Which is most likely to be the length of a football pitch?

    5 cm   100 mm   3 km   100 m

2. Which is most likely to be the length of a pencil?

    14 km   3 mm   14 cm   17 m

3. Which is most likely to be the mass of a newborn baby?

    1.5 g   4.5 g   3.8 t   3.5 kg

4. Which is most likely to be the capacity of a cup?

    14 *l*   0.5 m*l*   0.5 *l*   1.2 m*l*

5. Change these units of length.

    a) 2500 m = _____ km   b) 2.3 cm = _____ mm

    c) 35 mm = _____ cm   d) 350 cm = _____ m

    e) 7.5 km = _____ m   f) 8.9 m = _____ cm

    g) 3.74 cm = _____ mm   h) 760 m = _____ km

6. Change these units of mass.

    a) 15 kg = _____ g   b) 350 g = _____ kg

    c) 0.35 t = _____ kg   d) 0.45 kg = _____ g

    e) 6 t = _____ kg   f) 7400 g = _____ kg

    g) 150 kg = _____ g   h) 950 kg = _____ t

Revision Test on Length, Mass and Capacity *(continued)*

7. Change these units of capacity.

   a) $1000 \, ml = $ _____ l b) $3l = $ _____ ml

   c) $2.5l = $ _____ ml d) $2450 \, ml = $ _____ l

   e) $0.763 \, l = $ _____ ml f) $875 \, ml = $ _____ l

   g) $0.002 \, l = $ _____ ml

8. Michael wants to make some bookshelves. He has a piece of wood 3 m long. He wants to cut it into six pieces of equal length. How many centimetres long will each piece be?

9. Sarah has a mass of 38 kg. Her mass is ten times bigger than that of her baby sister. What is the mass of her baby sister in grams?

10. Mohan has 2.5 l of water in a big bottle. He pours in another 500 ml of water. How much water is there now in the bottle in litres? What is the mass of the water in the bottle now?

11. In a science experiment Isabelle adds 30 ml of a liquid to 0.75 l of another liquid. How much liquid does she have altogether in millilitres?

12. Jack is 1.38 m tall. He is 13 cm taller than his sister. How tall is his sister in metres?

10 Temperature and Time

Temperature

Temperature is the measure of how hot something is. We can measure it by using a **thermometer**. The type of thermometer we use most often is made from a glass tube with a liquid inside that rises up the glass tube as the temperature gets hotter.

Here is a drawing of a thermometer:

In the metric system we measure temperature in **degrees Celsius**. To save time we can write °C instead of writing degrees Celsius. So, for example, if the temperature is 20 degrees Celsius we can write 20 °C.

Measuring temperature in degrees Celsius was the idea of a scientist from Sweden called Anders Celsius. He used the temperatures at which water freezes and boils as the idea for his measuring system:

◆ 0 °C is the temperature at which water freezes.
◆ 100 °C is the temperature at which water boils.

In countries where it gets very cold the temperature can drop lower than freezing. We write temperatures lower than freezing as negative numbers. For example, if the temperature is 3 degrees below freezing we would write this as −3 °C.

Look at the drawing of the thermometer again. We can think of it as a number line running up and down instead of across the page. It is a number line showing both positive and negative numbers and we can use it in the same way as a normal number line. As we go up the thermometer the temperature gets hotter, and as we go down the thermometer the temperature gets colder.

If we want to know what the temperature would be if it started at 10 °C and went up by 13 °C, we add 13 to 10. We could do this in our heads, or we could use the thermometer as a number line to count on 13. The answer is 23, so the new temperature is 23 °C.

If we want to know what the temperature would be if it started at 20 °C and went down by 30 °C, we could subtract 30 from 20 in our head or we could count back down the number line. The answer is −10, so the new temperature would be −10 °C.

If the temperature starts at −3 °C and goes up by 7 °C, we could draw a number line and count on 7, or we could work out in our head that if we add 7 to −3 we will get to +4. So the answer is 4 °C.

Exercise 1

Now try these questions.

a) Which of these temperatures is the hottest?

   2 °C   5 °C   −12 °C   10 °C

b) Which of these temperatures is the coldest?

   3 °C   0 °C   −3 °C   5 °C

What is the new temperature if:

c) it starts at 2 °C and goes up by 3 °C

d) it starts at 2 °C and goes down by 3 °C

e) it starts at −1 °C and goes up by 2 °C

f) it starts at 5 °C and goes down by 5 °C

g) it starts at −3 °C and goes down by 4 °C

h) it starts at 5 °C and goes down by 8 °C

How do we work out how much the temperature has risen or fallen?

Sometimes we need to work out how many degrees the temperature has risen or fallen from one temperature to another. This is the **difference** between two temperatures.

For example, the temperature rises from −1 °C to 4 °C and we want to work out how much the temperature has gone up. We could draw our own number line or we could work out in our heads that from −1 °C to 0 °C is a rise of 1 °C, and then from 0 °C to 4 °C is another rise of 4 °C. If we add the rises together we will get how much the temperature has gone up altogether:

   $1 + 4 = 5$

The answer is 5 °C.

Exercise 2

a) If the temperature starts at 6 °C and goes up to 9 °C, how much has the temperature gone up?

b) If the temperature starts at 15 °C and goes up to 28 °C, how much has the temperature gone up?

c) The temperature starts at 24 °C and goes down to 19 °C. How much has the temperature gone down?

d) The temperature goes up from −2 °C to 1 °C. How much has the temperature gone up?

e) The temperature goes down from 7 °C to −2 °C. How much has the temperature gone down?

f) The temperature goes down from −4 °C to −8 °C. How much has the temperature gone down?

g) The temperature one morning was 18 °C. By lunchtime the temperature was 30 °C. How much had the temperature gone up?

h) One day at the North Pole the temperature went up from −21 °C to −16 °C. How much did the temperature go up?

Time

Time is measured in seconds, minutes, hours, days, weeks, months and years. We don't use a metric system to measure time so it doesn't have the same pattern as other measures.

Units of Time

The units we use for measuring time are:

```
60 seconds  = 1 minute
60 minutes  = 1 hour
24 hours    = 1 day
7 days      = 1 week
28–31 days  = 1 month
365 days    = 1 year (366 days in a leap year)
12 months   = 1 year
10 years    = 1 decade
100 years   = 1 century
```

The months of the year are:

January	31 days
February	28 days (29 days in a leap year)
March	31 days
April	30 days
May	31 days
June	30 days
July	31 days
August	31 days
September	30 days
October	31 days
November	30 days
December	31 days

A leap year happens once every four years.

Exercise 3

a) Which is most likely to be the time taken to eat lunch?

   8 seconds   2 seconds   30 weeks   35 minutes

b) Which is most likely to be the time taken to run 400 metres?

   10 years   3 months   1 minute   25 seconds

c) Which is most likely to be the time taken to blink your eyelids?

   12 weeks   1 year   1 second   2 days

d) Which is most likely to be the age of a grandmother?

   58 minutes   55 years   47 weeks   61 days

Changing Between Units

If we want to change units between seconds, minutes, hours, days and weeks we have to multiply or divide like this:

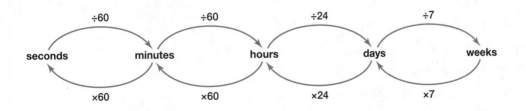

So, for example, if we want to change 2 minutes into seconds we multiply 2 by 60:

$2 \times 60 = 120$

So 2 minutes is 120 seconds.

If we want to change 360 minutes into hours we divide 360 by 60:

$360 \div 60 = 6$

So 360 minutes is 6 hours.

If we want to change 380 minutes into hours we divide 380 by 60:

$380 \div 60 = 6 \; r20$

The 20 remaining is 20 minutes so the answer is 6 hours and 20 minutes.

When answering questions about time we don't change the remainder to a decimal unless we are talking about fractions of a second.

Here is another example:

"Change 246 hours into days."

To change hours to days we divide by 24, so for this example we divide 246 by 24:

$246 \div 24 = 10 \; r6$

So the answer is 10 days and 6 hours.

Exercise 4

Complete these changes of units.

a) 180 seconds = _____ minutes

b) 72 hours = _____ days

c) 2 days = _____ hours

d) 5 weeks = _____ days

e) 5 minutes = _____ seconds

f) 240 seconds = _____ minutes

g) 21 days = _____ weeks

h) 10 days = _____ hours

i) 4 hours = _____ minutes

j) 100 minutes = _____ seconds

k) 1 hour and 10 minutes = _____ minutes

l) 3 weeks and 2 days = _____ days

m) 51 hours = _____ days and _____ hours

n) 37 days = _____ weeks and _____ days

Measuring Time

We measure time using a clock or a watch. Most clocks and watches show 12 hours, which means that each day the hour hand goes round the clock once between midnight and midday and then once between midday and midnight. The hour hand is the shorter hand. Midday is also called **noon**.

We read the hours round the clock as "1 o'clock", "2 o'clock", "3 o'clock" and so on all the way round to "12 o'clock". "O'clock" is a short way of saying "of the clock".

This is how 1 o'clock and 6 o'clock look on a 12-hour clock. The hour hand points to 1 at 1 o'clock and 6 at 6 o'clock.

The longer hand is the minute hand. This goes round the clock once every hour. But the number of minutes is not the same as the number that the long hand points to on the clock face. When the minute hand points straight upwards it is the start of the hour. So the minute hand points straight upwards at 1 o'clock, at 2 o'clock and so on.

There are 60 minutes in an hour so we must think of the clock face as being divided into 60 equal minutes. So when the minute hand is a quarter of the way round the clock face that is equal to 15 minutes past the hour (because 15 is one quarter of 60). When the minute hand is half way round the clock face that is equal to 30 minutes past the hour. When the minute hand is three-quarters of the way round the clock face that is equal to 45 minutes past the hour.

The 12 hours between midnight and midday are the morning hours and are called **a.m.** The 12 hours between midday and midnight are the afternoon and evening hours and are called **p.m.**

For example:

3 o'clock in the morning would be written as 3.00 a.m.
5 o'clock in the afternoon would be written as 5.00 p.m.
ten minutes past four o'clock in the afternoon would be written as 4.10 p.m.
20 minutes past midday would be written as 12.20 p.m.
5 minutes past 10 o'clock in the morning would be written as 10.05 a.m.

Exercise 5

Write these times as a.m. or p.m. times.

a) 6 o'clock in the morning

b) 4 o'clock in the afternoon

c) 20 minutes past 8 o'clock in the evening

d) 5 minutes past 3 o'clock in the morning

e) 10 minutes past 10 o'clock in the morning

f) 15 minutes past midday

The 24-hour Clock

We can also write the time of day using the 24-hour clock. When we write time using the 24-hour clock we don't use a.m. and p.m. This is because the 24-hour clock starts at midnight and each hour is counted from 1 to 12 as usual. But then at midday we don't go back to 1. We keep counting, so the next hour is 13, then 14, and so on all the way to midnight, which is 24. These 24-hour clock times are usually written with a colon (:) between the hours and the minutes.

For example:

1.00 p.m. is 13:00 as a 24-hour clock time

3.00 p.m. is 15:00 as a 24-hour clock time

Morning times stay the same, but we no longer need to write a.m. So 10.00 a.m. is 10:00 as a 24-hour clock time.

Here are some more examples:

3.15 a.m. is 03:15 as a 24-hour clock time

10.25 p.m. is 22:25 as a 24-hour clock time

6.45 a.m. is 06:45 as a 24-hour clock time

Notice than when the hour is less than 10 we normally put a 0 in front of it. For example 4.00 a.m. is written as 04:00.

Sometimes you will see 24-hour clock times written in other ways:

♦ Sometimes there is one dot instead of a colon between the hours and the minutes. So 15:25 would be written as 15.25.

♦ Sometimes the colon between the hours and the minutes is replaced with a space. So 15:25 would be written as 15 25.

♦ Another way is to put h for hours instead of the colon. So 15:25 would be written as 15h25.

Exercise 6

Write these times as 24-hour clock times.

a) 8.20 a.m.

b) 6.50 p.m.

c) 7.05 a.m.

d) Noon

e) 11.00 p.m.

f) 5.35 p.m.

Write these times as 12-hour clock times with a.m. or p.m.

g) 05:10

h) 21:00

i) 14:30

j) 12:45

k) 11:20

l) 01:15

Word Problems with Time

Adding and Subtracting Time

Sometimes we know when something starts and how long it lasts, and we then need to work out when it ends. For example:

"If it takes Rashid 45 minutes to get to work and he leaves his house at 7.10 a.m., what time will he get to work?"

To work out the answer we need to add 45 minutes to 7.10 a.m.

7.10 a.m. means 7 o'clock and 10 minutes, so if we add on 45 minutes to the 10 minutes that will be 55 minutes.

So Rashid will arrive at work at 7.55 a.m.

Sometimes we have to subtract or count backwards. Here is an example:

"Ruth took 35 minutes to get home from her friend's house. She arrived home at 18:05. What time did she leave her friend's house?"

This time we have to subtract or count backwards 35 minutes from 18:05.

5 minutes back from 18:05 would be 18:00.

30 minutes back from 18:00 would be 17:30.

So Ruth left her friend's house at 17:30.

Finding How Long Things Take

Sometimes we need to work out how long things take or how much time has passed between two things happening. For example:

"If Sarah left her house to go to school at 7.40 a.m. and arrived at school at 7.55 a.m. how long did it take her to get there?"

We can see that Sarah left home at 40 minutes past 7 o'clock and arrived at school at 55 minutes past 7 o'clock. So if we subtract 40 from 55, or count forward from 40 to 55, we will find the number of minutes it took her.

40 to 55 is 15

So it took Sarah 15 minutes to get to school.

Here is another example:

"A runner started a long-distance race at 13:45 and arrived at the finish at 15:10. How long did it take her to run the race?"

If we count forwards we can work out the answer:

from 13:45 to 14:00 is 15 minutes
from 14:00 to 15:00 is 60 minutes
from 15:00 to 15:10 is 10 minutes

Now we add the minutes together: 15 + 60 + 10 = 60 + 15 + 10 = 85 minutes

So the race took the runner 85 minutes, which we can write as 1 hour and 25 minutes because 60 of the minutes make up 1 hour.

Exercise 7

a) Adam left home to go to a football game at 2.10 pm. The game started 35 minutes later. What time did the game start?

b) The school play took 1 hour and 20 minutes. It finished at 19:25. What time did it start?

c) A lesson at school started at 10.40 a.m. and finished at 11.20 a.m. How long did the lesson last?

d) A plane took off at 09:20 and arrived at 11:00. How long did the flight last?

e) How long is it in minutes from 8.30 a.m. to 10.10 a.m.?

f) A radio programme started at 17:40 and ended at 18:25. How long was the programme?

Remember!

♦ We measure temperature in degrees Celsius.
♦ We write degrees Celsius as °C.
♦ Temperatures lower than freezing are written as negative numbers.
♦ There are 60 seconds in a minute, 60 minutes in an hour, 24 hours in a day and 7 days in a week.

Remember! *(continued)*

◆ When we use the 12-hour clock we write a.m. to show times in the morning and p.m. to show times in the afternoon and evening.

◆ When we use the 24-hour clock we usually add a colon (:) between the hours and the minutes, as in 10:25. Sometimes you might see this written as 10.25, 10 25, or 10h25.

Revision Test   on Temperature and Time

Now that you have worked your way through the chapter, try this revision test. The answers are in the answer book.

1. What is the new temperature if:
   a) it starts at 5 °C and goes up by 3 °C
   b) it starts at 2 °C and goes down by 7 °C
   c) it starts at −1 °C and goes up by 4 °C
   d) it starts at 35 °C and goes down by 8 °C
   e) it starts at −3 °C and goes down by 5 °C

2. If the temperature starts at 7 °C and goes up to 11 °C, how much has the temperature gone up?

3. If the temperature starts at 13 °C and goes up to 31 °C, how much has the temperature gone up?

4. The temperature rises from −3 °C to 2 °C. How much has the temperature gone up?

5. The temperature falls from 9 °C to −4 °C. How much has the temperature gone down?

6. When school started one morning the temperature was 18 °C. When school ended the temperature was 25 °C. How much had the temperature gone up?

7. Complete these changes of units.
   a) 300 seconds = _____ minutes
   b) 96 hours = _____ days
   c) 7 weeks = _____ days
   d) 15 minutes = _____ seconds
   e) 35 days = _____ weeks
   f) 20 days = _____ hours
   g) 4 hours = _____ minutes

Revision Test on Temperature and Time (continued)

    h) 2 hours and 15 minutes = _____ minutes

    i) 4 weeks and 3 days = _____ days

    j) 51 days = _____ weeks and _____ days

8. Write these times as a.m. or p.m. times.

    a) 9 o'clock in the morning

    b) 4 o'clock in the afternoon

    c) 15 minutes past 7 o'clock in the evening

    d) 35 minutes past 5 o'clock in the morning

9. Write these times as 24-hour clock times.

    a) 6.45 a.m.

    b) 9.35 p.m.

    c) 7.15 a.m.

    d) 11.10 p.m.

10. Write these times as 12-hour clock times with a.m. or p.m.

    a) 07:55

    b) 23:05

    c) 17:30

    d) 11:40

11. Jane left home with her parents to go into town at 09:50. They got to town 1 hour and 35 minutes later. What time did they get to town?

12. A bus journey took 2 hours and 20 minutes. It ended at 16:30. What time did the bus journey start?

13. A TV programme started at 7.15 p.m. and finished at 8.40 p.m. How long did the programme last?

14. Michael walked from home to his grandmother's house. He walked for 35 minutes and got to his grandmother's house at 2.20 p.m. What time did he leave home?

15. A ship sailed from one port to another. The ship started at 10:00 and ended the journey at 17:45. How long did the journey take?

11 Angles and Shapes

What Are Angles?

An **angle** is a measure of how much something changes direction or **turns**. For example, imagine you are standing at the centre of a circle with your arm pointing out in front of you like an arrow. If you turn one quarter of the way round the circle your arm will point in a different direction, like the arrows in this drawing:

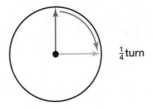

The angle you have turned is $\frac{1}{4}$ of the circle.

If you turn half way round the circle your arm will point in a different direction like this:

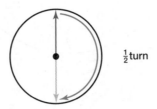

The angle you have turned is now $\frac{1}{2}$ of the circle.

If you turn all the way round the circle you will end up facing the same way you started:

We measure angles in **degrees**. These are not the same as the degrees we use to measure temperature. There are 360 degrees in a full turn. So if we turn all the way round a circle, and end up facing the same way we started, we have turned through 360 degrees. This is written as 360°.

If we turn half way round the circle we have turned through 180 degrees. This is because 180 is half of 360. 180 degrees is written as 180°. If we draw a half turn it gives us a straight line:

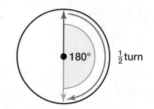

We can show the angles in a circle like this:

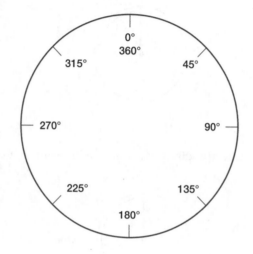

Types of Angles

Right Angles

A quarter turn is a 90° turn and we call this a **right angle**.

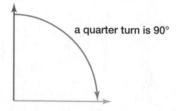

We show that an angle is a right angle by marking it with a small square like this:

90° is a right angle

Acute Angles

Any angle that is less than 90° is called an **acute angle**.

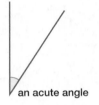

an acute angle

Obtuse Angles

Any angle that is between 90° and 180° is called an **obtuse angle**.

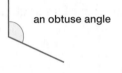

an obtuse angle

Reflex Angles

Any angle that is between 180° and 360° is called a **reflex angle**.

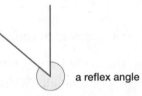

a reflex angle

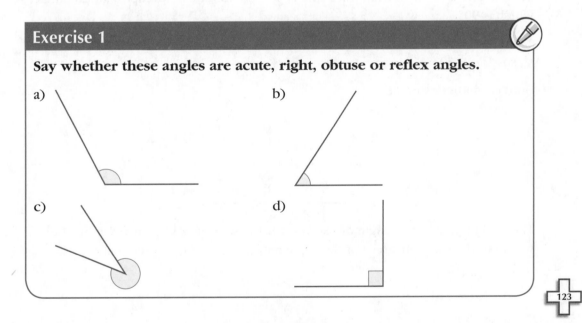

Exercise 1

Say whether these angles are acute, right, obtuse or reflex angles.

a)

b)

c)

d)

Working Out the Size of an Angle

We can measure angles using a **protractor**. A protractor looks like this:

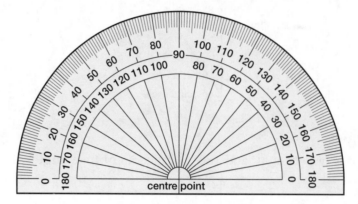

Sometimes we can work out the size of an angle from the facts we are given. Look at this example:

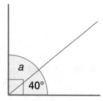

How do we work out the size in degrees of the angle marked as *a*?

The small square tells us that the angle between the two thick lines is a right angle. We know that there are 90° in a right angle. We are told that the angle between the thin line and the bottom thick line is 40°. So 40° and *a* add up to a right angle, which is 90°:

40° + *a* = 90°

so   *a* = 50°

Here is another example:

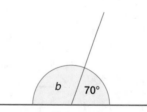

The straight line has an angle of 180° as it is the same as a half turn. We need to work out the size in degrees of the angle marked as *b*. We can see that:

70° + *b* = 180°

so   *b* = 110°

Here is another example:

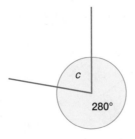

We know that a full turn is 360° so the angles must add up to 360°:

$$280° + c = 360°$$
$$\text{so} \quad c = 80°$$

Exercise 2

Work out the size in degrees of each angle marked with a letter.

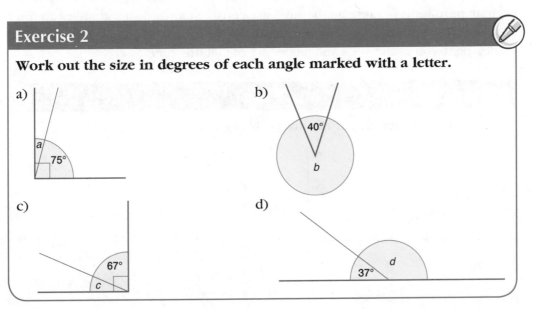

a)

b)

c)

d)

Shapes

Polygons

If the sides of a shape are all straight lines, then we call the shape a **polygon**.
Polygons with different numbers of sides have different names.

- ◆ A polygon with **3** sides is a **triangle**.
- ◆ A polygon with **4** sides is a **quadrilateral**.
- ◆ A polygon with **5** sides is a **pentagon**.
- ◆ A polygon with **6** sides is a **hexagon**.
- ◆ A polygon with **7** sides is a **heptagon**.
- ◆ A polygon with **8** sides is an **octagon**.

If all the sides of a polygon are equal length, and all the angles are the same size, then the polygon is called a **regular polygon**. Look at these two shapes:

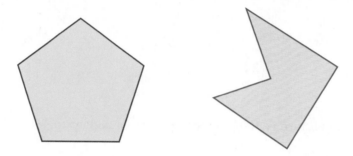

Both shapes are pentagons because they both have five sides. The first shape is a **regular pentagon** because all the sides are equal length and all the angles are the same size. The second shape is a **pentagon** because it has five sides, but it is not a regular pentagon as the sides and angles are not all the same.

Exercise 3

Name these shapes. Say if the shape is regular.

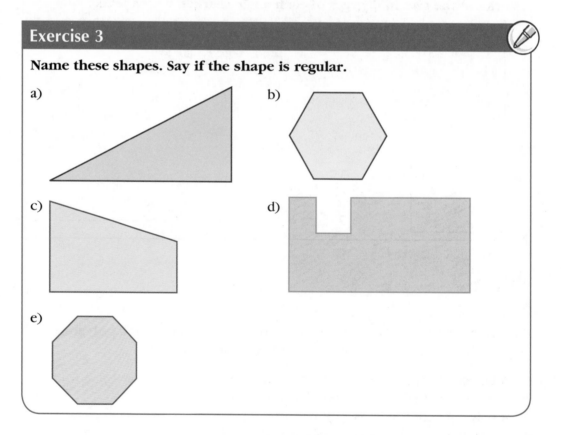

a)

b)

c)

d)

e)

Triangles

Triangles are polygons with three sides. There are four different types of triangle.

- ◆ **Equilateral**: if all three sides are the same length and all three angles are the same size then the triangle is an equilateral triangle.

- ◆ **Isosceles**: if two of the sides are the same length and two of the angles are the same size then the triangle is an isosceles triangle.

- ◆ **Scalene**: if all three sides are different lengths and all three angles are different sizes then the triangle is a scalene triangle.

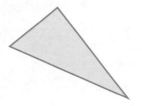

- ◆ **Right-angled**: if one of the angles is a right angle then the triangle is a right-angled triangle.

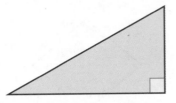

It is possible for both isosceles and scalene triangles to also have a right angle.

Quadrilaterals

The main quadrilaterals that we use have their own names. The quadrilaterals we use most often are in the list below.

♦ **Square**: all the sides are the same length and all the angles are right angles.

♦ **Parallelogram**: opposite sides are parallel and equal length. **Parallel** sides are straight lines that are exactly the same distance apart all along their length.

♦ **Rectangle**: opposite sides are parallel and equal length. All the angles are right angles. A rectangle is a special type of parallelogram.

♦ **Rhombus**: opposite sides are parallel and all the sides are the same length.

♦ **Trapezium**: one pair of sides is parallel. The other sides are not parallel.

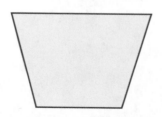

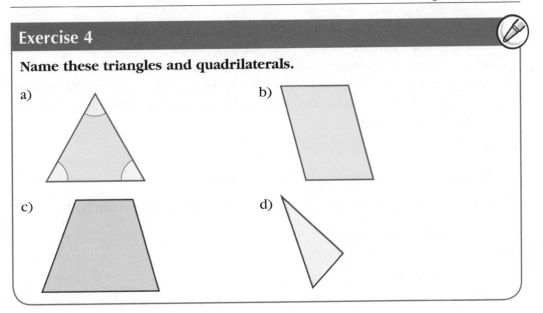

Exercise 4

Name these triangles and quadrilaterals.

a)

b)

c)

d)

Circles

The distance all the way round a circle is called the **circumference**. The distance from the outside of the circle to the centre of the circle is called the **radius**. The distance from one side of the circle through the centre to the other side of the circle is called the **diameter**.

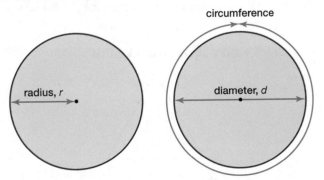

circumference

radius, *r*

diameter, *d*

We can see that the radius is half as long as the diameter. So if we know the length of the diameter we can divide it by 2 to get the length of the radius. If we know the length of the radius we can multiply it by 2 to get the length of the diameter.

So:

diameter = 2 × radius

And:

radius = diameter ÷ 2

So, for example, if the diameter = 6 cm then the radius = 3 cm.

If the radius = 4 cm then the diameter = 8 cm.

Many years ago people noticed that if you measure the circumference of a circle and then divide the circumference by the diameter of the circle you always get the same answer. It doesn't matter whether the circle is big or small. The circumference divided by the diameter always gives the same answer. The answer it gives, rounded to 2 decimal places, is 3.14. This is a very special number in mathematics. It is called **pi**.

So, for any circle:

circumference ÷ diameter = 3.14 (to 2 decimal places)

There is a special symbol for pi, which we write like this: . So we can write:

circumference ÷ diameter =

Congruent and Similar Shapes

Congruent means "the same". If two shapes are *exactly* the same size and shape we say they are **congruent shapes**. If we cut out one of the shapes and place it on top of the other shape it matches exactly. If two or more sides are the same length we say they are congruent sides. If two or more angles are the same size we say they are congruent angles.

For example, in an equilateral triangle all three sides are congruent because they are all the same length. All three angles are congruent because they are all the same size.

We mark congruent sides with a small dash like this:

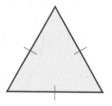

Similar shapes are the same shape but they are a different size. For example, these two shapes are similar, but they have different sizes so they are not congruent:

Symmetry

If we draw a line through the middle of a shape we can see whether the two halves of the shape are the same. We can check this by folding the shape along the line. If one half fits *exactly* on top of the other half, so that all the edges line up, then we know that the line is a **line of symmetry**. We normally show a line of symmetry with a dotted line like this:

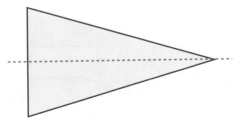

Some shapes have more than one line of symmetry. A square has four lines of symmetry as it can be folded along four different lines to give halves that match each other exactly. Here is a drawing of a regular hexagon showing three lines of symmetry. It can be folded exactly in half along these three different lines. Can you find any other lines of symmetry for this hexagon?

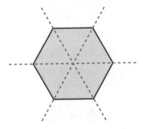

Polygons that aren't regular can also have lines of symmetry. This arrow has one line of symmetry:

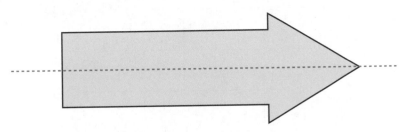

Exercise 5

Copy these shapes and draw the lines of symmetry.

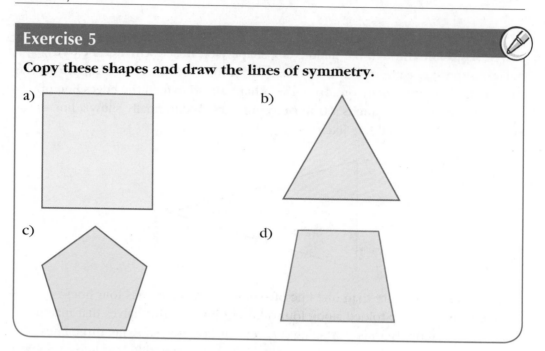

a)

b)

c)

d)

Tiling and Tessellation

We can use shapes to cover a flat place like a floor. This is called **tiling**. You may have seen tiling on a floor or on a wall. Normally we use congruent shapes for tiling. Remember that congruent shapes are shapes that are exactly the same. On floors and walls we often use congruent shapes for tiling with no gaps between them. When there are no gaps between the shapes, this is a special sort of tiling called **tessellation**.

Here is an example of tiling with congruent shapes that gives tessellation. There are no gaps between the shapes:

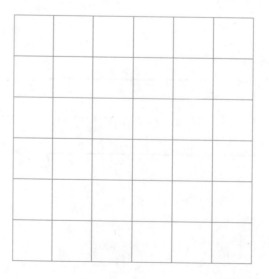

Here is an example of tiling with congruent shapes that doesn't give tessellation. There are gaps between the shapes:

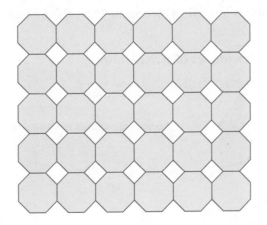

Solid Shapes

So far the shapes we have looked at are all flat like a piece of paper. A solid shape has thickness. We call the side of a solid shape a **face**. Solid shapes can have faces, edges and corners. Corners of solid shapes have a special name. We call the corners **vertices**.

Look at this drawing. This shape is called a **cube**.

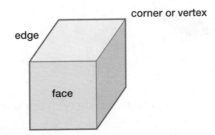

This shape has 6 faces, 12 edges and 8 corners or vertices.

Faces can be flat or curved. Edges can also be straight or curved. Faces meet at an edge, and edges meet at a corner or vertex.

Look at this drawing. This shape is called a **sphere**. It is the same shape as a ball, so a ball is a sphere.

A sphere has one curved face, no edges and no corners.

Polyhedrons

If all the faces of a solid shape are polygons then the solid shape is called a **polyhedron**. A cube is a polyhedron because all the faces are squares and a square is a polygon.

Here are some more examples of polyhedrons:

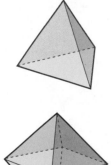

A polyhedron with 4 faces is called a tetrahedron.

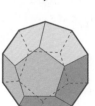

A polyhedron with 8 faces is called an octahedron.

A polyhedron with 12 faces is called a dodecahedron.

Prisms

If we have a solid shape that can be sliced from one end to the other into equal slices it is called a **prism**. For example, if we slice a cube, every slice will be exactly the same shape and size.

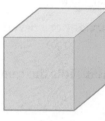

A cube is a prism.

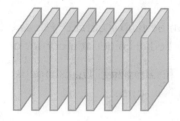

This shape is called a **cylinder**. It is also a prism because if we slice it from one end to the other every slice is exactly the same.

A cylinder is
a prism.

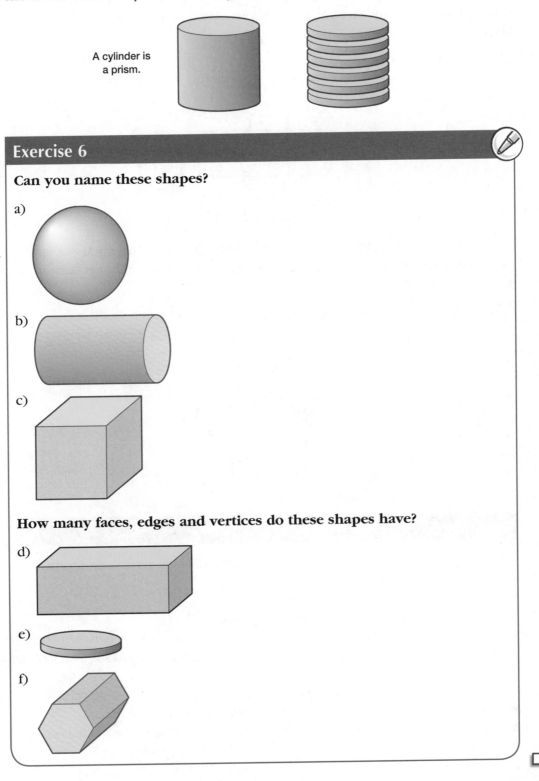

Exercise 6

Can you name these shapes?

a)

b)

c)

How many faces, edges and vertices do these shapes have?

d)

e)

f)

Nets

If we want to make a solid shape from a flat piece of paper or card we need to draw a **net** of the shape. We can then cut out the net of the shape and fold it to make a solid shape.

Here is a net for a cube:

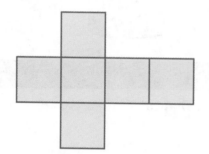

You could try copying this net onto a piece of paper. If you then cut it out and fold it along the lines you will make a cube. Try it! There are other nets that make a cube. Can you think of one?

Here is a net for a cylinder. If you cut out this shape and fold it you will make a cylinder:

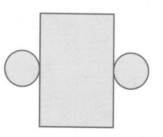

Remember!

- ◆ An angle is a measure of how much something changes direction or turns.
- ◆ We measure angles in degrees.
- ◆ There are 360° in a full turn or circle.
- ◆ An angle measuring 90° is called a right angle.
- ◆ Angles measuring less than 90° are called acute angles. Angles between 90° and 180° are called obtuse angles, and angles measuring between 180° and 360° are called reflex angles.
- ◆ If the sides of a shape are all straight lines, the shape is called a polygon.
- ◆ The distance all the way round a circle is called the circumference.
- ◆ The distance from the outside of the circle to the centre of the circle is called the radius.

Remember! *(continued)*

- ◆ The distance from one side of the circle through the centre to the other side of the circle is called the diameter.
- ◆ The circumference of a circle divided by the diameter always equals 3.14 (to 2 decimal places). This number is called "pi".
- ◆ If two shapes are exactly the same they are congruent shapes.
- ◆ If two shapes are the same shape but different sizes they are similar shapes.
- ◆ Solid shapes have faces, edges and vertices.
- ◆ If all the faces of a solid shape are polygons then the solid shape is called a polyhedron.
- ◆ If we want to make a solid shape from a flat piece of paper we need to draw a net of the solid shape.

Revision Test on Angles and Shapes

Now that you have worked your way through the chapter, try this revision test. The answers are in the answer book.

1. Are these angles acute angles, right angles, obtuse angles or reflex angles?

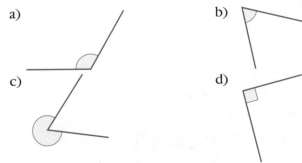

   a)

   b)

   c)

   d)

2. What is the size of the angle marked *a*?

3. What is the size of the angle marked *x*?

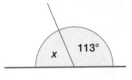

   →

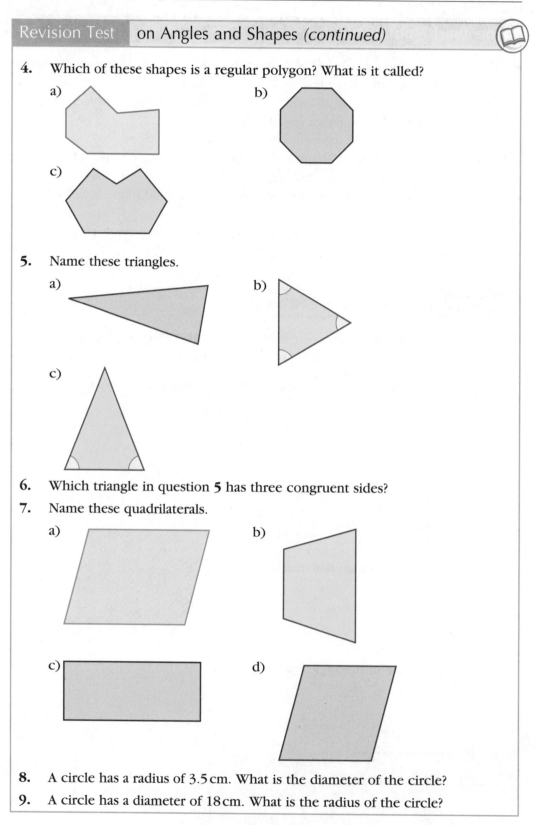

4. Which of these shapes is a regular polygon? What is it called?

a)

b)

c)

5. Name these triangles.

a)

b)

c)

6. Which triangle in question **5** has three congruent sides?

7. Name these quadrilaterals.

a)

b)

c)

d)

8. A circle has a radius of 3.5 cm. What is the diameter of the circle?

9. A circle has a diameter of 18 cm. What is the radius of the circle?

10. Copy these shapes and draw the lines of symmetry.

a)

b)

c)

d)

11. How many faces, edges and vertices do these shapes have?

a)

b)

c)

d)

12. Which solid shape goes with net A, net B and net C?

Net A

Net B

Net C

Solid shape 1

Solid shape 2

Solid shape 3

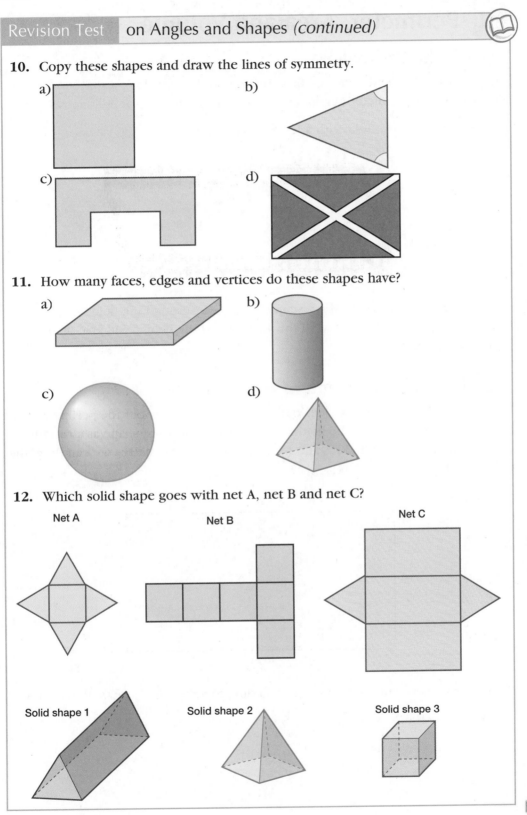

12 Perimeter, Area and Volume

What is the Perimeter?

The **perimeter** is the distance all the way around the outside of a space or a shape. We measure the perimeter in kilometres, metres, centimetres or millimetres.

Here is a drawing of a field with a fence around it:

The perimeter of the field is the distance all the way around the field. So if we measure the length of the fence all the way around the field it will give us the perimeter of the field.

Working Out the Length of the Perimeter

If we know the length of the sides of the field we can work out the perimeter by adding the sides together. The field drawn above is the same shape as a rectangle. If the long sides are 100 metres and the short sides are 50 metres we can draw the field like this:

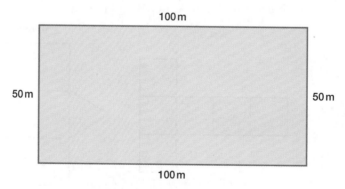

We can now work out the perimeter by adding together the lengths of the sides:

100 + 50 + 100 + 50 = 300

The perimeter is 300 metres.

Here is another example. Let's work out the perimeter of this rectangle:

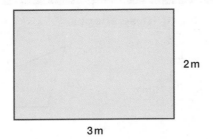

We know that the opposite sides of a rectangle are the same length, so it is easy for us to put in the lengths of the other two sides like this:

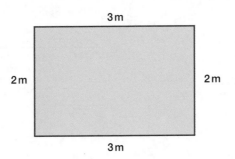

We can now add together all the sides to give us the perimeter:

$$3 + 2 + 3 + 2 = 10$$

So the perimeter is 10 metres.

Here is an example with a different shape. This time we want to work out the perimeter of a triangle:

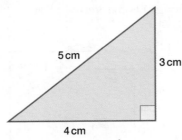

To work out the perimeter of the triangle we add together the lengths of the three sides:

$$3 + 4 + 5 = 12$$

The perimeter is 12 cm.

Exercise 1

Work out the perimeter of these shapes.

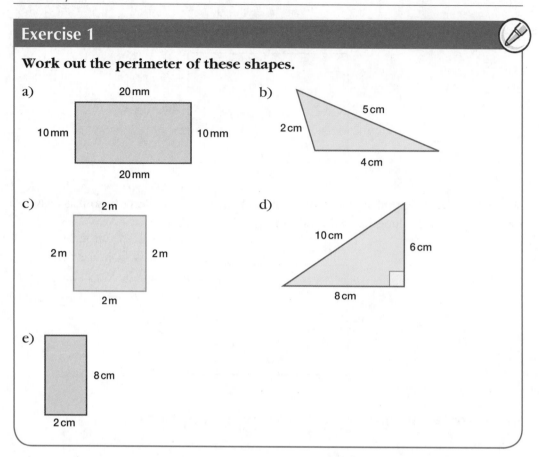

a) 20 mm / 10 mm / 10 mm / 20 mm

b) 5 cm / 2 cm / 4 cm

c) 2 m / 2 m / 2 m / 2 m

d) 10 cm / 6 cm / 8 cm

e) 8 cm / 2 cm

Sometimes we are not given all the lengths that we need to work out the perimeter, so we have to work out the missing lengths before we can add them all together.

In this shape we need to work out the missing length before we can work out the perimeter. The missing length is marked with an x:

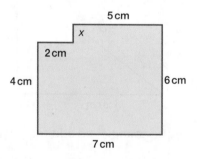

5 cm / x / 2 cm / 4 cm / 6 cm / 7 cm

We can see from the shape that $4 + x$ will be the same length as the opposite side of the shape, which is 6 cm:

$$4 + x = 6$$

So $x = 2$ cm.

We can now add together all the sides to give us the perimeter:

6 + 5 + 2 + 2 + 4 + 7 = 26

The total perimeter is 26 cm.

If a shape is drawn accurately we can measure the length of the sides using a ruler. **Accurately** means *exactly*. Here is an equilateral triangle drawn accurately:

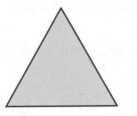

If we measure the sides using a ruler we can see that each side is 3 cm long, so the perimeter is 9 cm.

Exercise 2

Work out the perimeter of these shapes. None of the shapes is drawn accurately apart from shape e).

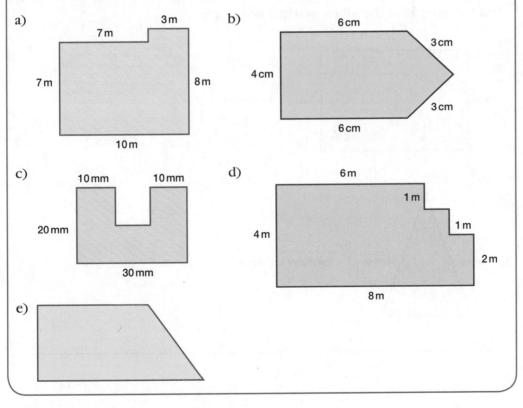

a)

7 m
3 m
7 m
8 m
10 m

b)

6 cm
3 cm
4 cm
6 cm
3 cm

c)

10 mm   10 mm
20 mm
30 mm

d)

6 m
1 m
1 m
4 m
2 m
8 m

e)

Sometimes we are told the perimeter of a shape and then we need to work out the length of a side. Here is an example:

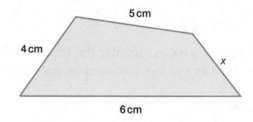

The perimeter of this shape is 22 cm. What is the length of the side marked with an x?

If we add together the lengths of the sides we can see that:

$$6 + 4 + 5 + x = 22$$
$$15 + x = 22$$
$$x = 7$$

So the length of side x is 7 cm.

Exercise 3

Find the length of the sides marked with an x.

a)

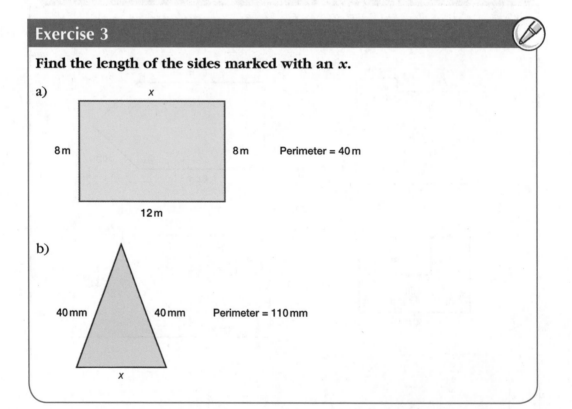

Exercise 3 *(continued)*

c)

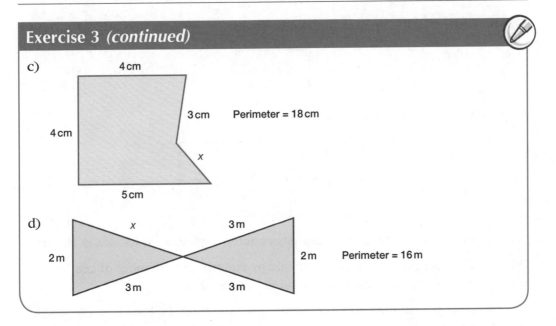

4 cm

3 cm   Perimeter = 18 cm

4 cm

x

5 cm

d)

x   3 m

2 m   2 m   Perimeter = 16 m

3 m   3 m

What is the Area of a Shape?

The **area** of a shape is the size of the shape. The size of the shape is how much flat space it covers.

Look again at our drawing of a field:

We know that the perimeter is the distance around the outside of the field. The area is the size of the field.

How Do We Measure Area?

Look at this square:

1 cm²

Each side of the square is 1 cm long, so this square is a centimetre square. This means that the area of the square is 1 square centimetre. We write this as 1 cm². The little "2" means *squared*.

This rectangle is made up of 2 centimetre squares, so its area is $2\,cm^2$.

Now have a look at this rectangle:

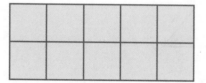

We can see that the rectangle covers 10 centimetre squares, so its area is $10\,cm^2$.

We can work out the area of the rectangle by counting the number of centimetre squares, but there is a quicker way to work out the area.

The rectangle is 5 cm long and 2 cm wide. If we multiply the length by the width we will get the area:

   $5\,cm \times 2\,cm = 10\,cm^2$

The area is $10\,cm^2$.

We can also find the area of a square by multiplying one side by the other. We can see by counting the squares that the area of this square is $9\,cm^2$:

If we multiply the length of one side by the length of the other we get the same answer:

   $3\,cm \times 3\,cm = 9\,cm^2$

The area is $9\,cm^2$.

So to work out the area of a rectangle or a square we multiply the length by the width:

   area = length × width

Not all shapes are measured in centimetres. Some shapes are smaller and are measured in millimetres. Some are much bigger and are measured in metres or even kilometres. So we must always look carefully at the units of length on drawings of shapes.

- ◆ If we measure a shape in millimetres its area will be measured in square millimetres, which we write as mm².
- ◆ If we measure a shape in centimetres its area will be measured in square centimetres, which we write as cm².
- ◆ If we measure a shape in metres its area will be measured in square metres, which we write as m².
- ◆ If we measure a shape in kilometres its area will be measured in square kilometres, which we write as km².

For example, remember our field:

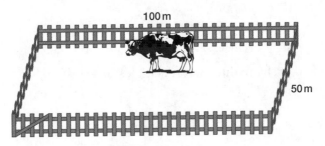

We said that the field is 100 m long and 50 m wide.

length × width = 100 × 50 = 5000

So the area of the field is 5000 m². (Remember that if we multiply something by 100 the digits move two places to the left, so 50 × 100 = 5000.)

Exercise 4

Find the area of these shapes.

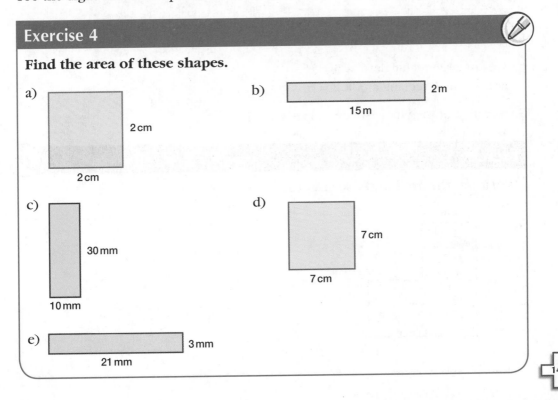

Sometimes the shape we are looking at is not a simple square or rectangle. We then need to split it up so that we can work out the area.

For example, how can we find the area of this shape?

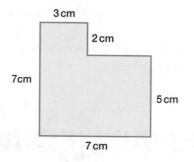

We can divide this shape into two rectangles like this:

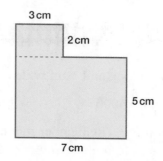

Now we can work out the area of both rectangles and add them together to give us the area of the whole shape.

area of large rectangle: $7 \times 5 = 35$
area of small rectangle: $3 \times 2 = 6$

So, the area of the whole shape = $35 + 6 = 41 \, cm^2$.

Exercise 5

Work out the area of these shapes.

a)

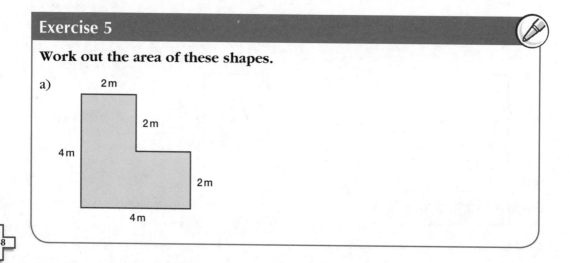

Exercise 5 *(continued)*

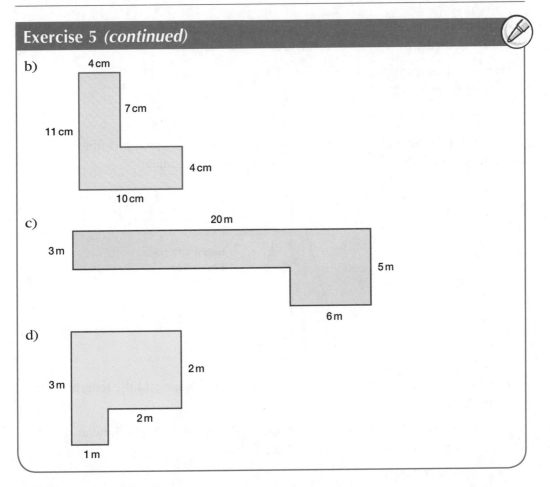

b)

c)

d)

Right-Angled Triangles

If we cut a rectangle in half from corner to corner we will have two right-angled triangles like this:

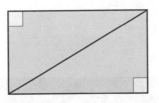

The area of each triangle is *half* the area of the rectangle. So we can work out the area of the triangle by multiplying the length by the width and then dividing by 2.

For example, look at this triangle. You can see from the broken lines that it is half of a rectangle:

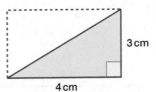

We multiply the length by the width to give us the area of the rectangle and then we need to divide by 2 to give us the area of the triangle:

area of rectangle: $4 \times 3 = 12$
area of triangle: $12 \div 2 = 6$

So the area of the triangle is $6\,\text{cm}^2$.

Working out the area in this way works for *any* triangle, even if it is not a right-angled triangle. For example, here is an isosceles triangle:

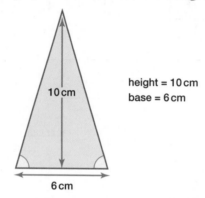

height = 10 cm
base = 6 cm

For this kind of triangle, we usually call the width the **base** and the length the **height**.

So for this triangle the base is 6 cm and the height is 10 cm. To find the area we multiply the base by the height and then divide by 2:

$$\begin{aligned}
\text{area} &= \frac{\text{base} \times \text{height}}{2} \\
&= \frac{6 \times 10}{2} \\
&= \frac{60}{2} \\
&= 30
\end{aligned}$$

The area is $30\,\text{cm}^2$.

Exercise 6

Work out the area of these triangles. The triangles are not drawn accurately.

a)   b)

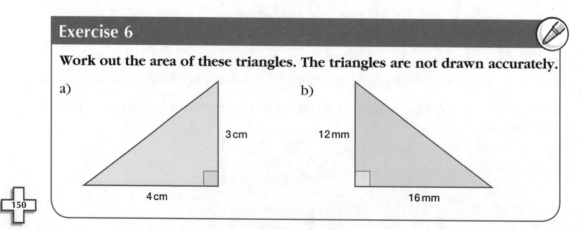

Exercise 6 *(continued)*

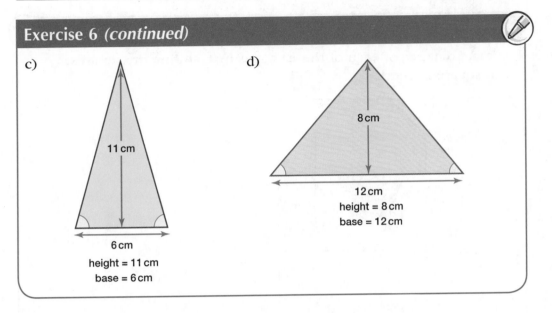

c)

11 cm

6 cm

height = 11 cm

base = 6 cm

d)

8 cm

12 cm

height = 8 cm

base = 12 cm

Estimating Area

Sometimes we might have to work out the area of an unusual shape as best we can without having accurate measurements. This is called **estimating** the area. For example, if we have a polygon like this and we want to work out the area, we can estimate:

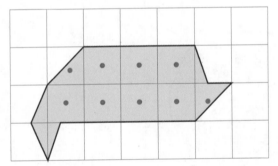

We estimate the area by counting the squares. If a square is half or more covered we count it as 1 square. If a square is less than half covered we don't count it. Each square in this drawing is $1\,cm^2$, so for this shape our estimate of the area will be $9\,cm^2$.

Exercise 7

Estimate the area of each of these shapes by counting the squares.
Each square is 1 cm².

a)

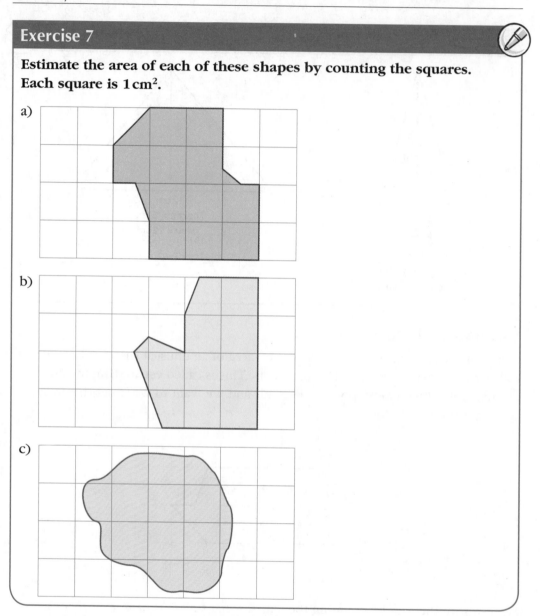

b)

c)

Area of a Circle

Remember from Chapter 11 *Angles and Shapes* that:

◆ the distance around a circle is called the circumference
◆ the distance from the edge of a circle to the centre is called the radius
◆ the distance across a circle from one side to the other through the centre is called the diameter.

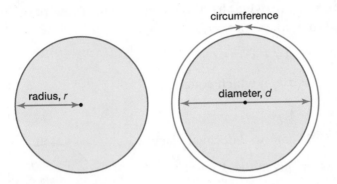

Remember also that if we divide the circumference of a circle by the diameter the answer is always the same. The answer is 3.14 to 2 decimal places. We call this answer "pi", which we show like this: π.

If we want to work out the area of a circle we first need to know the radius. We then need to multiply the radius by itself to give us the *radius squared*. Remember in Chapter 5 *More About Numbers* we saw that any number multiplied by itself gives a square number. We show a square number with a small "2". For example, $2 \times 2 = 2^2$, which equals 4.

We use the letter r for the radius, so the radius multiplied by itself is $r \times r$, which is r^2.

Once we know the value of r^2 we then need to multiply it by π to give us the area of the circle.

area of a circle $= \pi \times r^2$

So if the radius of a circle is 2 cm, the area of the circle is:

$\pi \times 2^2 = \pi \times 4$

We know that $\pi = 3.14$ (to 2 decimal places), so:

area $= 3.14 \times 4$
   $= 12.56$

The area of the circle is 12.56 cm^2 (to 2 decimal places).

You can work out 3.14×4 by using the method we learnt in Chapter 7 *Decimals*, or you can work it out on a calculator if you have one.

Here is another example:

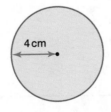

We can see from the drawing that the radius is 4 cm. To work out the area of the circle we need to work out $\pi \times r^2$:

area $= \pi \times r^2 = 3.14 \times 4^2$

We know that $4^2 = 4 \times 4$, which equals 16:

$3.14 \times 16 = 50.24$

So the area is $50.24 \, cm^2$ (to 2 decimal places).

In this circle we are given the diameter:

To find the area we need to work out $\pi \times r^2$.

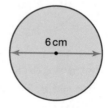

6 cm

Remember that the radius is half of the diameter. The diameter is 6 cm, so the radius must be 3 cm:

$\pi \times 3^2 = \pi \times 9 = 3.14 \times 9 = 28.26$

The area is $28.26 \, cm^2$ (to 2 decimals places).

Exercise 8

Work out the area of each of these circles. Use $\pi = 3.14$ and give your answers to 2 decimal places. You can use a calculator if you have one. If you don't have a calculator use the multiplication methods you learnt in Chapter 7 *Decimals*.

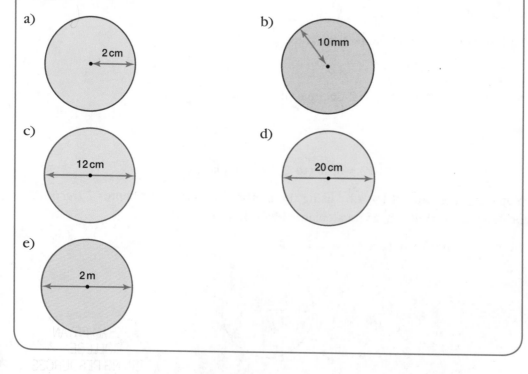

a)

2 cm

b)

10 mm

c)

12 cm

d)

20 cm

e)

2 m

Surface Area of a Solid Shape

Sometimes we want to work out the surface area of a solid shape. To do this we have to work out the area of each face and then add them all together.

For example, if we look at this cube we can see that each edge is 4 cm long.

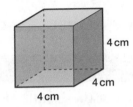

Each face is a square so we can work out the area of each face by multiplying the length by the width:

$4 \times 4 = 16$

So the area of each face is 16 cm². There are six faces altogether, and each face is the same:

$6 \times 16 = 96$

So the total surface area of the cube is 96 cm².

In this solid shape we can see that each face is a rectangle:

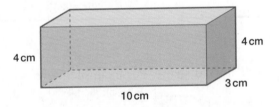

Two of the long faces (front and back) have the same area, 40 cm²:

$10 \times 4 = 40$

The other two long faces (top and bottom) have the same area, 30 cm²:

$10 \times 3 = 30$

So the total surface area of the long faces is 140 cm²:

$40 + 40 + 30 + 30 = 140$

The smaller faces at each end have the same area. The area of one of these faces is 12 cm²:

$4 \times 3 = 12$

So the area of the two smaller faces is 24 cm²:

2 × 12 = 24

Now we need to add the answers together to get the total surface area:

140 + 24 = 164

So the total surface area of the solid shape is 164 cm².

Exercise 9

Work out the surface area of these solid shapes.

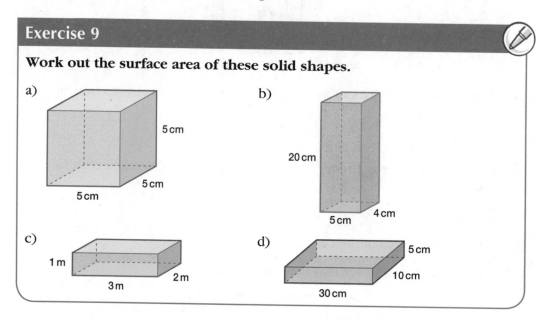

a)

5 cm

5 cm

5 cm

b)

20 cm

5 cm   4 cm

c)

1 m

3 m   2 m

d)

5 cm

10 cm

30 cm

What is Volume?

Volume is the space inside a solid shape. For example, here is a cube:

Volume = 1 cm³

Each edge of the cube is 1 cm long, so this cube is a centimetre cube. This means that the volume of the cube is 1 cubic centimetre. We write this as 1 cm³. The little "3" means *cubed*.

Each edge of this cube is 2 cm long:

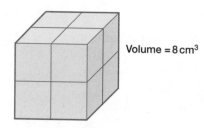

Volume = 8 cm³

We can see that the cube is made up from 8 cubic centimetres, so the volume of the cube is $8\,cm^3$.

This solid shape is 5 cm long, 2 cm wide and 2 cm high:

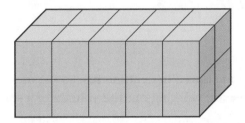

If we count the cubic centimetres we can see that there are 20 of them altogether, so the volume of this solid shape is $20\,cm^3$.

If the solid shape is a cube, or if every face of the solid shape is a rectangle (remember that a square is a type of rectangle), then we can find the volume by multiplying together the length, the width and the height.

volume = length × width × height

For example, if we look at this cube again we can count the cubic centimetres to see that the volume is $8\,cm^3$.

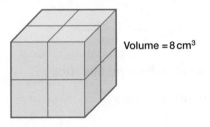

Volume $= 8\,cm^3$

But we can also multiply the length, the width and the height to give us the volume:

volume = length × width × height
    = 2 × 2 × 2
    = 8

So the volume is $8\,cm^3$.

Look at this solid shape:

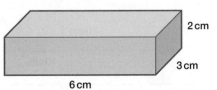

2 cm

3 cm

6 cm

Every face of the solid shape is a rectangle, so we can work out the volume of the solid shape by working out length × width × height:

volume = length × width × height
= 6 × 3 × 2
= 36

The volume is 36 cm³.

Not all solid shapes are measured in centimetres, so we must look carefully at the units of length when we are working out the volume of a solid shape.

♦ If we measure a shape in millimetres its volume will be measured in cubic millimetres, which we write as mm³.
♦ If we measure a shape in centimetres its volume will be measured in cubic centimetres, which we write as cm³.
♦ If we measure a shape in metres its volume will be measured in cubic metres, which we write as m³.

Here is an example of a solid shape measured in metres:

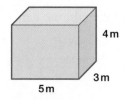

We work out the volume of this shape like this:

volume = length × width × height
= 5 × 3 × 4
= 60

So the volume is 60 m³. We would say this as "60 cubic metres".

Exercise 10

Work out the volume of these solid shapes.

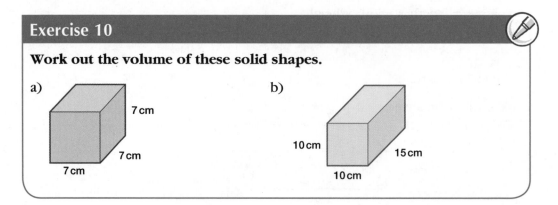

a)

b)

Exercise 10 *(continued)*

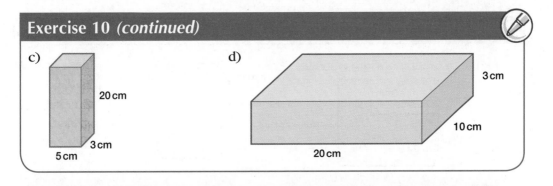

c)

20 cm

3 cm

5 cm

d)

3 cm

10 cm

20 cm

Remember!

◆ The perimeter is the distance all the way around the outside of a shape.
◆ The area of a shape is its size, which is the flat space that it covers.
◆ The volume is the space inside a solid shape.
◆ We usually measure area in square millimetres (mm²), square centimetres (cm²), square metres (m²) or square kilometres (km²).
◆ If the shape is a rectangle or a square, the area = length × width.
◆ The area of a triangle $= \dfrac{\text{base} \times \text{height}}{2}$
◆ The area of a circle $= \pi \times r^2$.
◆ We can estimate the area of unusual shapes by counting squares. We count only those squares that are half full or more.
◆ We can work out the surface area of a solid shape by working out the area of each face and then adding them together.
◆ We usually measure volume in cubic millimetres (mm³), cubic centimetres (cm³) or cubic metres (m³).
◆ If the solid shape is a cube, or if every face of the solid shape is a rectangle, then the volume = length × width × height.

Revision Test | on Perimeter, Area and Volume

Now that you have worked your way through the chapter, try this revision test. The answers are in the answer book.

1. Work out the perimeter of each of these shapes.

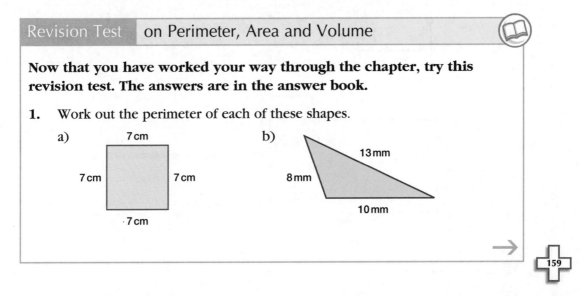

a)

7 cm

7 cm   7 cm

7 cm

b)

13 mm

8 mm

10 mm

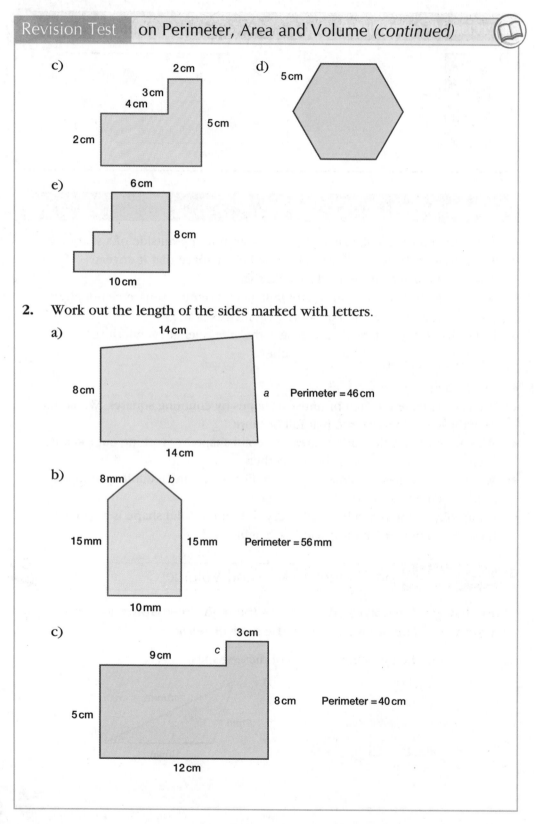

c)

2 cm

3 cm
4 cm

5 cm

2 cm

d)

5 cm

e)

6 cm

8 cm

10 cm

2. Work out the length of the sides marked with letters.

a)

14 cm

8 cm

a   Perimeter = 46 cm

14 cm

b)

8 mm   *b*

15 mm   15 mm   Perimeter = 56 mm

10 mm

c)

3 cm

9 cm   *c*

8 cm   Perimeter = 40 cm

5 cm

12 cm

Revision Test on Perimeter, Area and Volume *(continued)*

d)

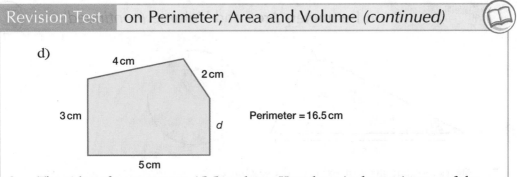

4 cm
2 cm
3 cm   Perimeter = 16.5 cm
          d
5 cm

3. The sides of a square are 15.5 cm long. How long is the perimeter of the square?

4. The perimeter of an equilateral triangle is 210 mm. How long is each side?

5. Estimate the area of each of these shapes. Each square is 1 cm².

a)

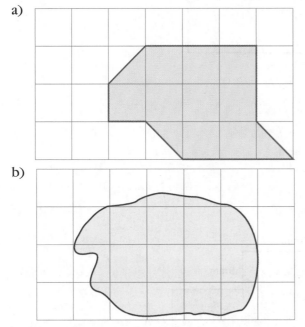

b)

6. Work out the area of each of these shapes.

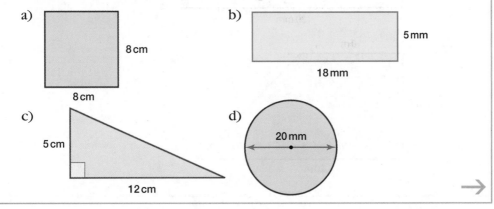

a)

8 cm

8 cm

b)

5 mm

18 mm

c)

5 cm

12 cm

d)

20 mm

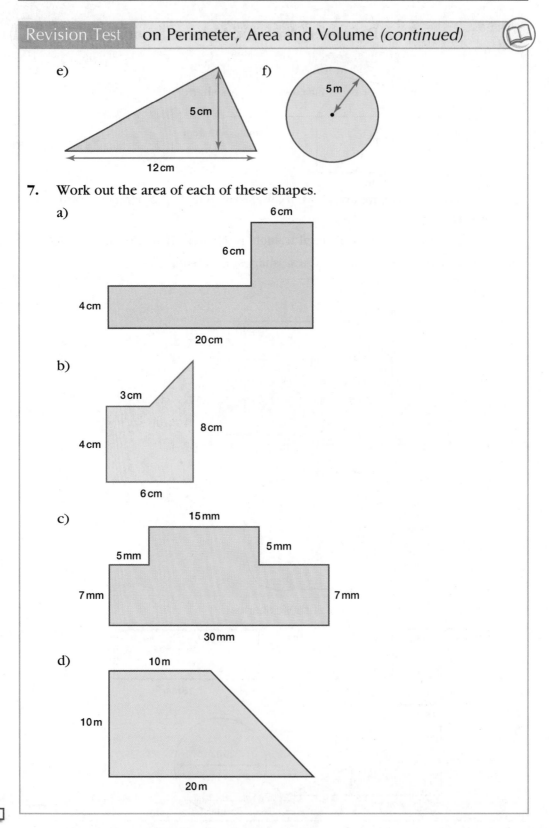

e)

5 cm

12 cm

f)

5 m

7. Work out the area of each of these shapes.

a)

6 cm

6 cm

4 cm

20 cm

b)

3 cm

8 cm

4 cm

6 cm

c)

15 mm

5 mm

5 mm

7 mm

7 mm

30 mm

d)

10 m

10 m

20 m

8. A swimming pool is 15 m long and 5 m wide. What is the area of the pool?

9. Work out the area of the shaded part of this shape.

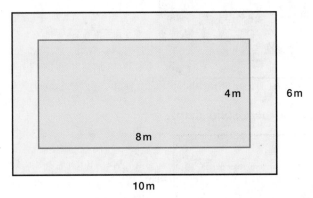

10. A circle has a diameter of 12 mm. What is the area of the circle?

11. Work out the total surface area of each of these solid shapes.

a)

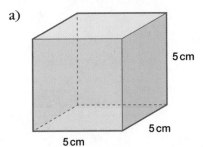

b)

c)

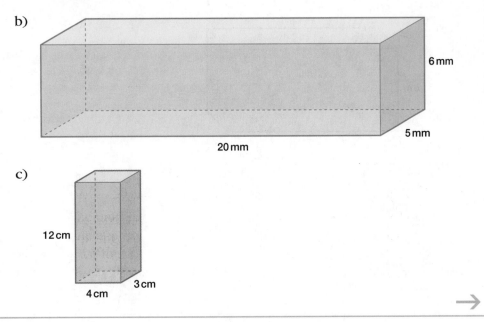

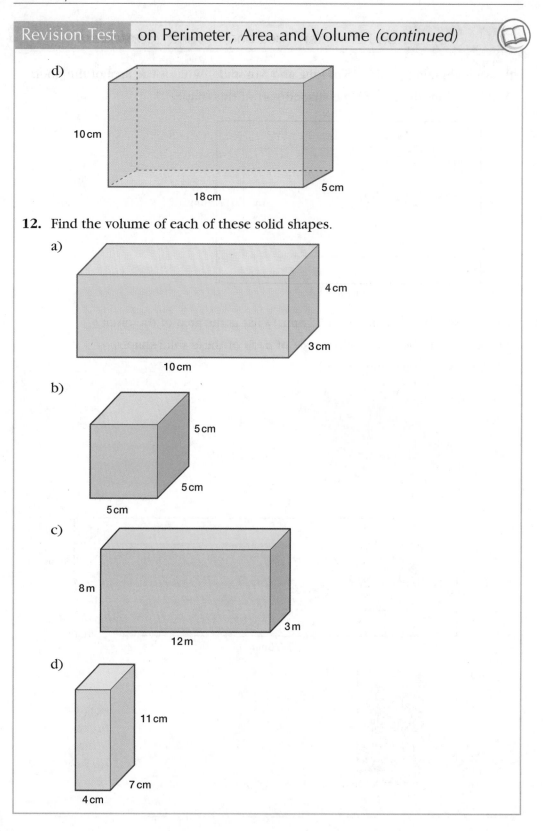

d)

10 cm

18 cm

5 cm

12. Find the volume of each of these solid shapes.

a)

4 cm

3 cm

10 cm

b)

5 cm

5 cm

5 cm

c)

8 m

3 m

12 m

d)

11 cm

7 cm

4 cm

Revision Test on Perimeter, Area and Volume *(continued)*

13. What is the volume of a cube with edges that are 4 cm long?

14. A storeroom is 5 m long, 3 m wide and 3 m high. What is the volume of the storeroom in cubic metres?

15. A cube has edges that are 3 m long. A rectangular prism is 4 m long, 3 m wide and 2 m high. Which of these two shapes has the largest volume? What is the difference in volume between the two shapes?

Statistics – Collecting and Working with Data

13 Collecting and Showing Data

What is Data?

Pieces of information are often called **data**. For example, all these things are examples of data:

◆ how many matches the school football team wins
◆ how many people like drinking milk
◆ how many people prefer swimming to running.

Collecting Data

A simple way of getting, or **collecting**, data is to ask people questions and write down their answers. If we want the data to be useful then we must make sure that we ask good questions that tell us what we need to know. When we are asking questions we should follow these rules:

◆ We should have a clear idea what it is we are trying to find out.
◆ We should ask questions that are clear and easy to understand.
◆ We should only ask questions that are about the thing we are trying to find out.
◆ Our questions should give people more than two possible answers to choose from whenever possible.
◆ Our questions should not be **biased**. This means that our questions should not make people give one answer instead of another.
◆ We should never ask questions that upset or embarrass people.

We must also remember to ask as many people as we can. The more people we ask the better our results will be. The number of people we ask is called the **sample**. To get useful results we need to make the sample as big as possible.

We often write down the data we collect in a table. Look at the example on page 167. 20 children were asked which is their favourite sport. The answers were recorded in the table.

Favourite sport	Number of children
running	6
swimming	3
football	5
netball	4
cricket	1
no favourite	1

You can see from the table that 6 children said running is their favourite sport, 3 said swimming, 5 said football, 4 said netball, 1 said cricket and 1 said they don't have a favourite sport.

Displaying Data

Once we have collected our data we need to find a way of showing it so that we can work out what it means. Showing data in a clear way is often called **displaying** data. There are lots of different ways of displaying data.

Block Graphs

Look again at the table above showing children's favourite sports. We can show this data in a block graph. In the block graph we draw a block for each child. So for running there are 6 blocks because 6 children said it is their favourite sport. For swimming, there are 3 blocks, and so on.

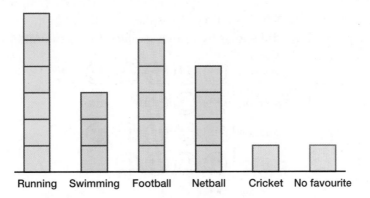

Bar Graphs

In a bar graph we don't draw separate blocks to show how many of something we have counted. Instead we draw a scale alongside the graph that gives us the numbers counted. We need to make sure that we make the scale a good size for showing the data without taking up too much or too little space. For our data on favourite sports the bar graph would look like this:

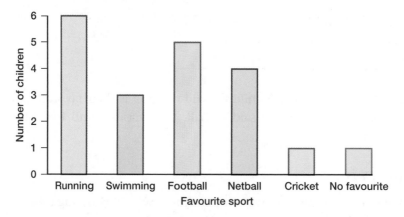

The scale gives the number of children. The height of each bar tells us how many children chose that sport as their favourite. For example, if we look across to the scale we can see that the top of the bar for netball is level with the 4 on the scale, because 4 children chose netball as their favourite sport. The top of the bar for football is level with the 5 on the scale because 5 children chose football as their favourite sport.

Pictographs

In a pictograph we draw small pictures to display our data. For example, we could display our data on favourite sports by drawing a face for each child like this:

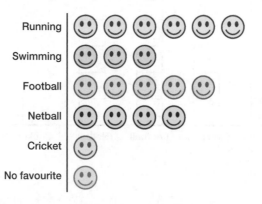

Swimming has 3 faces because 3 children chose swimming as their favourite sport. Cricket has 1 face because 1 child chose cricket, and so on. The pictures you choose to draw in a pictograph should all be the same size and should be spaced equally.

Pie Charts

Pie charts are often a good way of showing how data is divided between things, but they are sometimes a little more difficult to draw than other graphs. Pie charts are circles that are divided up to show data. We can show our data in a pie chart with each sport covering a different amount of the circle. The amount of the circle a sport covers tells us how many children chose that sport.

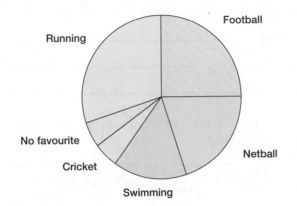

20 children were asked to name their favourite sport, so the size of our sample is 20. The whole circle therefore represents all 20 children. We know that there are 360° in a circle (if you have forgotten this look back at Chapter 11, *Angles and Shapes*). So 20 children take up 360°. To find out how many degrees 1 child takes up in the circle we need to divide 360° by 20.

$$360° \div 20 = 18°$$

Cricket and "no favourite" each take up 18° of the circle as each of these was chosen by 1 child.

Swimming was chosen by 3 children, so it takes up $3 \times 18° = 54°$ of the circle.

Netball was chosen by 4 children, so it takes up $4 \times 18° = 72°$ of the circle.

Football was chosen by 5 children, so it takes up $5 \times 18° = 90°$ of the circle.

Running was chosen by 6 children, so it takes up $6 \times 18° = 108°$ of the circle.

We can check if we have got the number of degrees right by adding them all together. They should add up to 360° to complete the circle.

$$18° + 18° + 54° + 72° + 90° + 108° = 360°$$

Exercise 1

a) 20 pupils were asked in a survey how they get from home to school.
    This table gives the data from the survey. Show this data as a bar graph.

How pupils travel to school	Number of pupils
walk	9
bicycle	5
car	2
bus	4

b) Here is a table showing the number of girls in each primary class at a school.
    Show this data as a bar graph.

Primary class	Number of girls
Class 1	18
Class 2	15
Class 3	12
Class 4	19
Class 5	11
Class 6	10

c) 12 people were asked at an airport which country they had come from. This
    table gives the results of the survey. Show this data as a pie chart.

Country	Number of people
Ghana	4
Jamaica	3
Nigeria	2
United Kingdom	2
USA	1

Line Graphs

We use line graphs when we want to show data that changes over time.
For example, we might have data that shows how far someone runs over an
hour-long period.

David ran 10 kilometres in one hour. The distance he had run was recorded every
ten minutes. Here is a table showing the results.

Time after start (in minutes)	10	20	30	40	50	60
Distance run (in km)	2	4	6	7	8	10

We can show this data on a line graph. First we show the distance in kilometres up
the side of the graph, then we show the time that passes along the bottom of the
graph. Time always goes along the bottom of a line graph.

Next we take each piece of data and mark it on the graph with a dot or a small
cross. For example, after 10 minutes David had run 2 kilometres, so we find "10
minutes" along the bottom of the graph and "2 km" up the side of the graph. We
run our finger or a ruler along from each of these numbers, and at the point where
they cross we mark our dot or small cross. When we have done this for each piece
of data in the table we can join them all together to give a line graph like this:

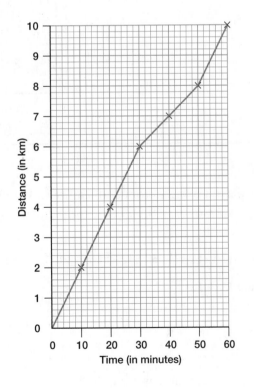

Exercise 2

a) In an experiment, Rosa heated a liquid. She measured the temperature of the liquid every five minutes for 20 minutes. This table gives her results. Show this data as a line graph.

Time after start (in minutes)	0	5	10	15	20
Temperature (in °C)	21	30	38	42	46

b) Look at this line graph of the distance cycled by Tom in 30 minutes.

How far had Tom cycled after 10 minutes?

c) Look at the line graph of Tom's cycle ride again. How long had Tom been cycling when he reached 6 km?

Stem and Leaf Plots

Sometimes data is displayed using a stem and leaf plot. A stem and leaf plot is a useful way of showing a lot of numbers. For example, here are the numbers of pupils in 14 primary schools:

123, 126, 137, 115, 118, 123, 125, 132, 116, 134, 116, 124, 139, 128

We can show this data like this:

```
11 | 5   6   6   8
12 | 3   3   4   5   6   8
13 | 2   4   7   9
```

In this stem and leaf plot, 11|5 means 115.

Showing large amounts of data in a stem and leaf plot saves space and makes it easier to see what numbers we have in our data. For example, if we want to know how many primary schools have between 120 and 129 pupils we can see from the stem and leaf plot that the answer is 6 schools.

Here is another example of 20 numbers displayed in a stem and leaf plot:

12, 15, 26, 14, 19, 22, 23, 31, 15, 35, 20, 19, 13, 27, 15, 33, 17, 11, 39, 24

```
1 | 1  2  3  4  5  5  5  7  9  9
2 | 0  2  3  4  6  7
3 | 1  3  5  9
```

Exercise 3

a) Show these 15 numbers in a stem and leaf plot.

   12, 25, 14, 17, 18, 30, 32, 21, 17, 15, 26, 33, 31, 26, 22

b) Here are the numbers of runs scored by 10 cricket teams in matches played one Saturday.

   145, 153, 167, 154, 141, 150, 162, 168, 154, 142

   Show these numbers in a stem and leaf plot.

c) In a town, the temperature was measured (in °C) at midday for 20 days. Here are the results.

   22, 24, 23, 29, 31, 25, 20, 19, 18, 22, 27, 26, 28, 31, 29, 34, 21, 24, 25, 26

   Show the data in a stem and leaf plot.

Remember!

◆ Pieces of information are often called data.
◆ We can collect data by asking questions.
◆ When collecting data, the sample needs to be as large as possible to get useful results.
◆ Data can be written down in a table.
◆ We can display data using graphs.
◆ We can use different types of graphs to display data, such as block graphs, bar graphs, pictographs or pie charts.
◆ We can use line graphs to show data that is measured over a period of time.
◆ Stem and leaf plots are useful for displaying a lot of numbers.

Revision Test on Collecting and Showing Data

Now that you have worked your way through the chapter, try this revision test. The answers are in the answer book.

1. Here is a bar graph showing the rainfall each month in London for one year.

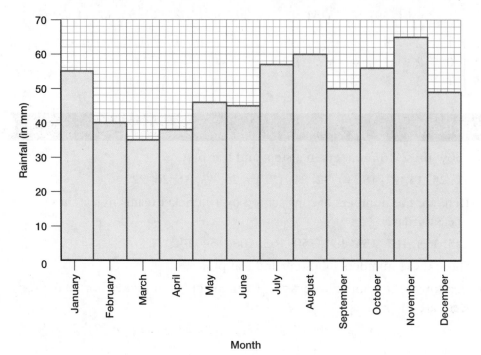

a) Which was the wettest month?

b) Which was the driest month?

c) What was the rainfall in March?

d) What was the rainfall in September?

2. 20 people were asked to name their favourite animal. Here are the results.

Animal	Number of people
cat	4
dog	6
rabbit	2
donkey	3
horse	5

Show these results in a bar graph.

Revision Test on Collecting and Showing Data (continued)

3. Here is a line graph showing the temperature in a schoolyard between 8.00 a.m. and 4.00 p.m.

a) What was the temperature at 10.00 a.m.?

b) What was the temperature at 2.00 p.m.?

c) What was the highest temperature between 8.00 a.m. and 4.00 p.m.?

d) How much did the temperature increase between 9.00 a.m. and 11.00 a.m.?

4. Here is a table showing the number of hours John spent doing different things in one day.

Show this data as a pie chart.

Activity	Number of hours
sleeping	10
at school	7
eating	2
playing	3
reading	2

5. A survey was taken of the number of people on 20 buses leaving the city centre. Here are the results.

52, 42, 39, 54, 43, 47, 40, 42, 46, 51, 52, 38, 37, 45, 47, 51, 38, 48, 49, 53

a) Show the data in a stem and leaf plot.

b) How many buses had fewer than 50 passengers?

c) How many buses had 50 or more passengers?

14 Averages and Probability

What is an Average Number?

If we have lots of data, it is useful to work out *one* number that can tell us something about all the data we have. The number we work out is called an **average**.

We can **compare** one set of data with another set of data very easily if we have an average number for each set. *Comparing* data means looking at one set of data next to another set of data. For example, we compare data if we want to see which set of numbers is bigger or smaller.

There are three kinds of average, called the **mean**, the **median** and the **mode**. Each one tells us something different about a set of numbers. For each kind of average we need to know what it means and how we work it out.

Working Out Averages

The Mean

The **mean** is the average we use most often in mathematics. When newspapers, TV and radio talk about an "average", they are usually talking about the mean.

We can work out the mean of a set of numbers by adding up all the numbers and then dividing the total by the number of numbers in our data. For example, here are four numbers:

2, 6, 5, 3

If we add them all together we get:

2 + 6 + 5 + 3 = 16

There are four numbers so next we divide the total by 4:

16 ÷ 4 = 4

So the mean for the set of numbers is 4.

Here is another example. 10 pupils score these marks in a spelling test:

8, 7, 8, 9, 7, 6, 5, 7, 10, 3

What is the mean score?

First we add up all the scores:

8 + 7 + 8 + 9 + 7 + 6 + 5 + 7 + 10 + 3 = 70

Next we divide the total by the number of scores:

70 ÷ 10 = 7

So the mean score in the spelling test was 7.

Mean numbers are not always whole numbers. For example, here is another set of 10 scores from a quiz:

4, 3, 5, 6, 4, 2, 1, 0, 5, 4

If we add these numbers together we get:

$4 + 3 + 5 + 6 + 4 + 2 + 1 + 0 + 5 + 4 = 34$

Now if we divide the total by the number of scores, we get:

$34 \div 10 = 3.4$

So the mean score is 3.4.

The Median

The **median** is the number in the middle of our data. We find the median by putting all the numbers in order from smallest to biggest. The median is the number in the middle of the list.

Here is a list of five numbers:

8, 4, 6, 5, 2

If we put these numbers in order from smallest to biggest we get:

2, 4, 5, 6, 8

We can see that the number in the middle of the list is 5. So the median is 5.

If there are two numbers in the middle, then the median is half way between the two middle numbers. For example, here is a list of six marks from a Science test:

7, 4, 6, 3, 8, 9

If we put these marks in order from smallest to biggest we get:

3, 4, 6, 7, 8, 9

We can see that 6 and 7 are both equally in the middle of this list, so the median is half way between these two numbers. Half way between 6 and 7 is 6.5. So the median is 6.5.

If the two numbers equally in the middle are the same, then the median is that number. For example, look at this list of numbers, which are in order from smallest to biggest:

2, 3, 4, 4, 6, 8

The two numbers in the middle are both 4, so the median is 4.

The Mode

The **mode** is the number that appears most often in a set of numbers.

For example, in this set of numbers we can see that 3 is the mode as there are more 3s than any other number:

2, 3, 7, 5, 3, 0, 2, 8, 1, 3

Normally, the mode is an average that is only useful when we have a very large sample, which means that there are lots of numbers in our data.

Exercise 1

a) Here are the goals that a football team scored in 10 matches.

1, 3, 3, 5, 0, 2, 1, 3, 0, 3

Find the mean, median and mode of the football team's scores.

b) What is the mean of these four numbers?

2.5, 3.0, 4.4, 2.1

c) Here are the marks out of 25 that six pupils got in a singing competition. What was the median mark?

18, 24, 23, 20, 17, 21

d) Here are the percentages that Joseph got in his end-of-year tests.

Maths 74%

Science 76%

English 60%

What was Joseph's mean percentage for these three tests?

e) Anil keeps chickens. He counted the number of eggs laid by the chickens each day for seven days. Here are the results.

4, 2, 4, 5, 3, 5, 5

What were the mean, median and mode of the number of eggs laid?

The Range

The **range** is not an average but it is a number we often use with averages. The range tells us the **spread** of a set of numbers from the smallest number to the biggest number. The range is very easy to work out. All we have to do is subtract the smallest number from the biggest number.

range = biggest number − smallest number

Look at this set of data:

18, 16, 19, 12, 17, 14, 15, 10, 14, 15

If we put these numbers in order from smallest to biggest, we get:

10, 12, 14, 14, 15, 15, 16, 17, 18, 19

We can see that the smallest number is 10 and the biggest number is 19.

To find the range we subtract the smallest number from the biggest number:

range = 19 − 10
     = 9

So the range of the numbers is 9.

Here is an example using temperatures. Here are the temperatures (in °C) recorded in a village at midday every day for seven days:

21, 26, 23, 23, 22, 27, 24

If we put these in order we get:

21, 22, 23, 23, 24, 26, 27

To find the range we subtract the lowest temperature from the highest:

27 − 21 = 6

So the range is 6 °C.

Finding the Range, Median and Mode from a Stem and Leaf Plot

You will remember from Chapter 13 *Collecting and Showing Data* that a stem and leaf plot is a useful way of displaying a lot of numbers. It is also easy to find the range, median and mode from a stem and leaf plot.

For example, the number of children born in a city hospital was counted every month for two years. Here are the results:

15, 21, 17, 28, 25, 30, 31, 32, 16, 22, 25, 23,
33, 20, 27, 21, 25, 19, 17, 23, 32, 18, 14, 16

If we put this data into a stem and leaf plot it looks like this:

1	4	5	6	6	7	7	8	9	
2	0	1	1	2	3	3	5	5	5	7	8
3	0	1	2	2	3				

In this plot, 1| 4 is 14.

We can find the range by subtracting the smallest number from the biggest number:

33 − 14 = 19

We can find the median by finding the middle number. 22 and 23 are equally in the middle, so the median is 22.5.

We can find the mode by looking for the number that appears most often. The answer is 25.

Exercise 2

a) Lisa runs 100 metres and times how long it takes. She does this five times. Here are her times in seconds.

14.0, 13.5, 12.5, 13.0, 13.0

What are the range, mean and median of her times?

b) What are the range and the mode of these ten numbers?

193, 162, 187, 210, 175, 175, 187, 201, 154, 187

c) In a test a class got these marks out of 40.

21, 25, 23, 19, 18, 35, 33, 31, 30, 20,

25, 28, 27, 26, 18, 28, 19, 17, 24, 28

i) Put the marks into a stem and leaf plot.

ii) What is the range of these marks?

iii) What is the median mark?

iv) What is the mode?

d) There are 10 boxes of pencils in the school cupboard. The teacher counts the number of pencils in each box. Here are the results.

16, 21, 19, 17, 18, 18, 21, 17, 18, 16

i) What is the mean number of pencils in a box?

ii) What is the median number of pencils in a box?

iii) What is the mode?

Probability

Probability is another word for *chance*. Probability is a measure of *how likely* something is to happen.

For example, we know that if a baby is about to be born it will be either a boy or a girl. We say that there are 2 possible *outcomes*, boy or girl. An **outcome** is something that can happen.

The baby being a boy is 1 outcome out of 2 possible outcomes. So there is a *1 in 2 chance* of the baby being a boy.

We write this as $\frac{1}{2}$. This is the probability of the baby being a boy.

The baby being a girl is 1 outcome out of 2 possible outcomes. So there is a *1 in 2 chance* of the baby being a girl.

We write this as $\frac{1}{2}$. This is the probability of the baby being a girl.

If we toss a coin, there are 2 possible outcomes. The coin can land on one side or the other.

The two sides are often called "heads" and "tails". They are equally likely, so the probability of the coin landing as "heads" is $\frac{1}{2}$ and the probability of the coin landing as "tails" is also $\frac{1}{2}$. This means that the probability of the coin landing as "heads" is a 1 in 2 chance, and the probability of the coin landing as "tails" is a 1 in 2 chance.

If we roll a die like the one in the drawing there are 6 possible outcomes:

1, 2, 3, 4, 5, 6

Each number has an equal chance of being thrown. So each number has a 1 in 6 chance of being thrown. We write this as $\frac{1}{6}$.

If we want to work out the probability of an even number being thrown on a die we first need to count up the number of even numbers. There are three even numbers on a die:

2, 4 and 6

So there are 3 possible outcomes that will give us an even number.

There are 6 possible outcomes altogether: 1, 2, 3, 4, 5 and 6.

So the probability of an even number being thrown is 3 chances out of 6. We write this as $\frac{3}{6}$.

You will remember from Chapter 6 *Fractions and Ratios* that $\frac{3}{6} = \frac{1}{2}$.

So the probability of an even number being thrown is $\frac{1}{2}$.

We can work out the probability of a result happening by following this rule:

$$\text{probability} = \frac{\text{number of possible outcomes that give the result}}{\text{number of all possible outcomes}}$$

Here is another example.

> Hannah puts four balls in a bag. Three of the balls are black and one is white. Adam closes his eyes and pulls one of the balls out of the bag. What is the probability of the ball being black?

To work this out we need to see that if there are 4 balls in the bag and 3 of them are black, then the probability of pulling out a black ball is 3 out of 4. We write this as $\frac{3}{4}$. We can also see that the probability of pulling the white ball out of the bag is 1 out of 4, which we write as $\frac{1}{4}$.

Sometimes you will see probability written as a decimal. For example, if a result has a probability of $\frac{1}{2}$ it will sometimes be written as 0.5. A probability of $\frac{1}{4}$ can be written as 0.25. If something is certain to happen, it has a probability of 1. If something is impossible, it has a probability of 0.

Exercise 3

a) Ruth tosses a coin. What is the probability of it landing as "tails"?

b) Saul puts five bottle tops in bag. Three of them are coloured silver and two of them are coloured gold. He asks his friend to pick one out of the bag without looking. What is the probability of the bottle top being silver?

c) Anna throws a die. What is the probability of the number being 5?

d) Anna throws the die again. What is the probability of the number being odd?

Remember!

◆ Averages help us compare data.

◆ There are three kinds of average, called the mean, the median and the mode.

◆ We work out the mean by adding together all the numbers in the data and then dividing the total by the number of numbers in the data.

◆ The median is the number in the middle of the data after all the numbers have been ordered from smallest to biggest.

◆ The mode is the number that appears most often in the data.

◆ The range is the smallest number subtracted from the biggest number.

◆ Probability is another word for chance.

◆ We can find the probability of an event happening by working out the number of possible outcomes that give the event and then dividing that number by the total number of all possible outcomes.

◆ Probability is usually written as a fraction, but it can also be written as a decimal.

Revision Test on Averages and Probability

Now that you have worked your way through the chapter, try this revision test. The answers are in the answer book.

1. A nurse measured the mass of six children to the nearest kilogram. Here are the results.

   40 kg, 35 kg, 42 kg, 38 kg, 45 kg, 46 kg

   a) What is the mean of these masses?

   b) What is the range of masses?

   c) What is the median mass?

2. A scientist measured the temperature in degrees Celsius at midnight for 10 days. Here are the results.

   10, 14, 12, 6, 11, 9, 14, 8, 9, 14

   a) What was the mean temperature?

   b) What was the median temperature?

   c) What was the mode?

   d) What was the range of temperatures?

3. Lesedi scored these marks in his English tests in year 6.

   80%, 75%, 60%, 70%, 58%

   a) What was Lesedi's mean percentage?

   b) What was Lesedi's median percentage?

   c) What was the range of Lesedi's marks?

4. Mary threw a die. What is the probability of the number being:

   a) 5

   b) 2

   c) odd

   d) even?

5. There are eight pencils in a box. Three of the pencils are red, two of the pencils are green, one of the pencils is blue and two of the pencils are yellow. A pencil is pulled out of the box without looking.

   a) What is the probability of the pencil being:

      i) blue

      ii) red

      iii) green

      iv) yellow?

   b) i) Which colour is most likely to be pulled from the box?

      ii) Which colour is least likely to be pulled from the box?

Algebra

15 Working with Algebra

What is Algebra?

Sometimes we see letters in mathematical sentences. This can make the mathematics look difficult, but it is often very easy.

For example, if we have 3 books and we add another 2 books, we then have 5 books:

3 books + 2 books = 5 books

Instead of writing the word "books" we could use the letter b. Our sentence would then be:

$3b + 2b = 5b$

In this sentence b means "books".

Instead of b meaning the word *books*, we can use b to mean *number of books*. Using letters to mean *number of something* is called **algebra**.

$2b$ then means 2 × *the number of books*
$3b$ then means 3 × *the number of books*

And so on. If there are 2 books then b equals 2. So $4b$ means $4 ×$ *the number of books*, which now equals $4 × 2 = 8$.

Adding and Subtracting with Algebra

Here is another example of a mathematical sentence:

3 boys + 2 boys + 4 girls = 5 boys and 4 girls

If we use b for "boys" and g for "girls" we get:

$3b + 2b + 4g = 5b + 4g$

Exercise 1

Complete these mathematical sentences in which the letters are used to represent objects. Here is an example.

3 cars + 4 bicycles + 2 cars = 5 cars + 4 bicycles
$3c + 4b + 2c = 5c + 4b$

a) 6 cows + 5 cows = _____ cows

$6c + 5c =$ _____ c

Exercise 1 *(continued)*

b) 4 bananas + 8 bananas = _____ bananas

$4b + 8b =$ _____ b

c) 5 girls − 3 girls = _____ girls

$5g - 3g =$ _____ g

d) 3 shirts + 2 trousers + 4 shirts = _____ shirts + _____ trousers

$3s + 2t + 4s =$ _____ $s +$ _____ t

e) 5 pencils + _____ pencils = 9 pencils

$5p +$ _____ $p = 9p$

In algebra, we add and subtract letters in the same way. Here are some examples:

$b + b = 2b$ $4b - b + 2g = 3b + 2g$

$3c + 2c = 5c$ $8g - 3g = 5g$

But in algebra we must remember that the letters mean *number of something*, like *number of boys* or *number of girls* or *number of cars*. Here are some more examples. In these examples, t means *number of trucks*.

$t + t = 2t$

$7t + 3t = 10t$

$5t - 2t = 3t$

Exercise 2

Complete the following.

a) $3a + 7a =$

b) $14p - 8p =$

c) $a + a =$

d) $4r + 3r + 6s =$

e) $2x + x =$

f) $5a + 6b + 8a =$

g) $8x + 4x + 3y =$

h) $3n + 4n + 2n + n =$

i) $2p + 3q + 7p - q =$

j) $45a - 12a + 17b - 4b =$

k) $12m - 3n + 17m + 9n =$

l) $-2a + 3b + 6c + 5a - c =$

Multiplying and Dividing with Algebra

We can also use multiplication and division in algebra.

For example, we write $2 \times a$ as $2a$ and $5 \times b$ as $5b$.

We would write $a \times b$ as ab and $m \times n$ as mn.

We would write $a \div 2$ as $\dfrac{a}{2}$ and $x \div 4$ as $\dfrac{x}{4}$.

We would write $a \div b$ as $\dfrac{a}{b}$ and $x \div y$ as $\dfrac{x}{y}$.

Here are a few more examples. They all follow the same rules, whatever letters we use:

$$3 \times y = 3y$$

$$7 \times x = 7x$$

$$c \times d = cd$$

$$2c \times d = 2cd$$

$$b \div 3 = \dfrac{b}{3}$$

$$m \div n = \dfrac{m}{n}$$

$$2a \div b = \dfrac{2a}{b}$$

Exercise 3

Complete the following. First, here is an example. You can also look at the examples above to help you.

$$7 \times s = 7s$$

a) $8 \times a =$ \qquad   b) $11 \times b =$

c) $x \times y =$ \qquad   d) $x \div y =$

e) $d \div 2 =$ \qquad   f) $r \times s =$

g) $3a \times b =$ \qquad   h) $3a \times 2b =$

i) $2x \div y =$ \qquad   j) $a \div 2b =$

Using Algebra to Find Unknown Values

We can use algebra to find the value of missing numbers. Algebra is a very useful way of helping us to find the value of numbers we don't know. We call numbers we don't know *unknown* values.

For example, look at this mathematical sentence:

6 sweets − x sweets = 4 sweets

In this simple sentence we need to work out the value of x. If we can work out the value of x it will tell us how many sweets we subtract from 6 sweets to give us 4 sweets. We can write this sentence like this:

$6 - x = 4$

We can see that $x = 2$ because $6 - 2 = 4$.

Here is another example:

$8 + y = 12$

What is the value of y? We can see that $y = 4$ because $8 + 4 = 12$.

We can also find unknown values in multiplication and division questions, like this example:

$4n = 16$

What does n equal?

We know that $4n$ means $4 \times n$:

$4 \times n = 16$
so $n = 4$

n must equal 4, because $4 \times 4 = 16$.

Here is another example:

$18 \div n = 6$

What does n equal?

We know that $18 \div 3 = 6$, so n equals 3.

Exercise 4

What does x equal in each of the following?

a) $4 + x = 6$

b) $x + 8 = 12$

c) $5 + x = 6$

d) $x + 50 = 100$

e) $9 - x = 4$

f) $12 - x = 2$

g) $57 - x = 40$

h) $3 \times x = 6$

i) $5x = 25$

j) $3x = 21$

k) $15 \div x = 5$

l) $\dfrac{x}{10} = 2.5$

187

Using Algebra with Given Values

Another way we use algebra is to find the answers to questions when we are given values to use. For example, look at this mathematical sentence:

$x + y$

That doesn't tell us anything on its own, but if we are asked to find out how much it equals if $x = 3$ and $y = 2$ we can see that $x + y$ becomes:

$x + y = 3 + 2 = 5$

Here is another example:

$2a - b$

What does this equal if $a = 6$ and $b = 3$?

First we must put in the values we are given:

$$2a - b = 12 - 3$$
$$= 9$$

So $2a - b = 9$, if $a = 6$ and $b = 3$.

Exercise 5

Use these values to complete the following.

$a = 5, b = 4$

a) $a + b =$

b) $a - b =$

c) $a \times b =$

d) $a^2 =$

e) $2a - b =$

f) $8 + a - b =$

g) $18 - a =$

h) $a + b^2 - 6 =$

i) $b \div 2 =$

j) $2a \times b =$

k) $3a \div 5 =$

l) $a^2 - b =$

Remember!

- We can use letters in mathematical sentences.
- In algebra, letters are used to mean number of objects or they can be used to represent values.
- There are simple rules for writing algebra.
- $a + b$ is written as $a + b$.
- $a - b$ is written as $a - b$.
- $a \times b$ is written as ab.
- $2 \times a$ is written as $2a$.

Remember! *(continued)*

◆ $a \div 2$ is written as $\dfrac{a}{2}$.

◆ These rules are true for all letters and numbers.
◆ We can use algebra to find unknown values.
◆ We can use algebra to work out answers when we are given values.

Revision Test on Working with Algebra

Now that you have worked your way through the chapter, try this revision test. The answers are in the answer book.

1. 19 fish + 6 fish = _____ fish

2. $19f + 6f =$ _____ f

3. 15 boys + 3 boys + 12 girls = _____ boys + _____ girls

4. $15b + 3b + 12g =$

5. $4a + 7a =$

6. $12x - x + 5y =$

7. $8m + 6n - 3m + 2n =$

8. $25s - 13s =$

9. Work out the value of a in each of the following.

   a) $4 + a = 28$ b) $19 - a = 11$

   c) $6 + 2a = 12$ d) $-a + 16 = 14$

   e) $4a = 40$ f) $6a + 3 = 45$

   g) $56 \div a = 7$ h) $a^2 = 81$

   i) $\dfrac{a}{2} = 8$ j) $\dfrac{45}{a} = 15$

10. Use these values to answer these questions.

    $x = 2, y = 3, z = 4$

    a) $x + y + z =$ b) $2x + z =$

    c) $\dfrac{z}{2} + y =$ d) $3y - z =$

    e) $2x + 2y + 2z =$ f) $x \times y =$

    g) $\dfrac{4x}{z} =$ h) $x^2 + y - z =$

    i) $-x + z =$ j) $2xy + z =$

Money

16 Money

Currency

All around the world, money is used to pay people for their work and to buy things. But the money used in one country is not always the same as the money used in another country. The money used in a country is called its **currency**. For example, the money used in Jamaica is the dollar. So the currency of Jamaica is the dollar. Most currencies have a special sign. The sign for the dollar is $.

Here is a list of some of the currencies used around the world:

Country	Name of currency	Sign
Antigua & Barbuda	Dollar	$
Barbados	Dollar	$
Belgium	Euro	€
Botswana	Pula	P
Egypt	Pound	£
France	Euro	€
Germany	Euro	€
Ghana	Cedi	₵
India	Rupee	Rs
Indonesia	Rupiah	Rp
Jamaica	Dollar	$
Kenya	Shilling	Sh
Malawi	Kwacha	K
Malaysia	Ringgit	RM
Mozambique	Metical	MTn
Nigeria	Naira	₦
Rwanda	Franc	RWF
Singapore	Dollar	$
South Africa	Rand	R

Country	Name of currency	Sign
St Lucia	Dollar	$
Tanzania	Shilling	Sh
Trinidad & Tobago	Dollar	$
Uganda	Shilling	Sh
United Kingdom	Pound	£
United States of America	Dollar	$
Zambia	Kwacha	K

You can see from the table that lots of countries use the dollar for their currency. But the dollar in one country is not always the same as the dollar in another country. For example, the dollars that are used in Singapore are not the same as the dollars that are used in Barbados.

Units of Money

Currencies are usually divided into 100 units. For example, the 100 units that make up dollars are called cents. So

100 cents = 1 dollar

Other currencies are also often made up of 100 units. For example:

◆ Pounds used in the UK are made up of pence. 100 pence = 1 pound
◆ Cedis used in Ghana are made up of pesewa. 100 pesewa = 1 cedi
◆ Naira used in Nigeria are made up of kobo. 100 kobo = 1 naira
◆ Ringgit used in Malaysia are made up of sen. 100 sen = 1 ringgit
◆ Euros used in Germany are made up of cents. 100 cents = 1 euro
◆ Rands used in South Africa are made up of cents. 100 cents = 1 rand

We can write 100 cents equals 1 dollar like this:

100¢ = $1

So 200 cents would be 2 dollars:

200¢ = $2

Because there are 100 cents in 1 dollar, we can think of 50 cents as being half a dollar. We write this as $0.50. We put a zero after the 5 to remind us that we are talking about 50 cents. 150 cents is one and a half dollars which we write as $1.50.

We write other currencies in exactly the same way. For example, we can think of 150 pence as being one and a half pounds, which we write as £1.50.

Exercise 1

Write down these amounts in dollars. For example, 145 cents = $1.45.

a) 85 cents

b) 30 cents

c) 5 cents

d) 62 cents

e) 125 cents

f) 345 cents

g) 840 cents

h) 104 cents

Working Out Problems with Money

Money is usually based on the decimal system. We add, subtract, multiply and divide money in exactly the same way that we add, subtract, multiply and divide decimals. If you are not sure how to work with decimals look again at Chapter 7 *Decimals* in this book.

We can work out how much change we get in a shop by subtracting one amount of money from another amount. For example, Luke bought some food in a shop for $4.20. He gave the shopkeeper $5.00. How much change does he get?

To work this out we need to find the difference between $4.20 and $5.00, so we subtract 4.20 from 5.00:

$$
\begin{array}{r}
{}^{4}5\ .\ {}^{1}0\ \ 0 \\
-\ 4\ .\ 2\ \ 0 \\
\hline
0\ .\ 8\ \ 0
\end{array}
$$

The answer is 0.80 so Luke gets $0.80 in change. We can also work this out by counting on from 4.20 to 5.00.

Multiplying an amount of money is useful for working out how much we need to pay to buy more than one of something. For example, if a shirt costs $50.00 and we buy three shirts, we can work out how much we pay by multiplying 50.00 by 3:

50.00 × 3 = 150.00

So 3 shirts cost $150.00.

If we want to share money we can use division. For example, if we have $450.00 and we want to share it equally between 5 people, we divide 450.00 by 5:

450.00 ÷ 5 = 90.00

So each person would get $90.00.

Percentages are also very useful when working out problems with money. When we are shopping we sometimes see signs giving a **discount**. A discount means that the price is lower than the normal price. A discount is usually given as a percentage. The following example of a discount uses pounds as the currency. We would work it out the same way in any other currency.

A notebook in a shop normally costs £8.00. The shop is selling it at 20% discount. What is the new price of the book?

To work this out we find 20% of £8.00. If you can't remember how to do this, look back at Chapter 8 *Percentages*.

The quickest way to work this out is to remember that 10% of 8.00 is 0.80, so 20% of 8.00 will be 1.60. So the discount is £1.60.

The new price will be

£8.00 − £1.60 = £6.40

A longer way to work this out is to remember that 20% is $\frac{20}{100}$. We then work out 20% of 8 like this:

$$8 \times \frac{20}{100} = \frac{8}{1} \times \frac{20}{100} = \frac{160}{100} = 1.60$$

So the discount is £1.60 and the new price is £8.00 − £1.60 = £6.40.

Another way to work out the answer is to remember that if the price is 20% less than the normal price, we will pay 80% of the normal price. So we can find 80% of the normal price:

10% of £8.00 = £0.80, so 80% of £8.00 must be 8 × £0.80 = £6.40.

So we pay £6.40.

Another use for percentages with money is to work out **interest**. Interest is the amount of money we get if we put our money in a bank. For example, if a bank offers 5% interest per year on the money we put in the bank, we can work out how much interest we get if we put $500.00 in the bank.

10% of 500.00 is 50.00

So 5% of 500.00 is 25.00

So the interest we get in a year is $25.00.

We can also work this out like this:

$$500 \times \frac{5}{100} = \frac{500}{1} \times \frac{5}{100} = \frac{2500}{100} = 25$$

So the interest we get is $25.00.

Exercise 2

Find the answer to these questions on money.

a) Jane buys a shirt, a dress and some shoes. The shirt costs $75.00. The dress costs $45.00 and the shoes cost $110.00. How much does Jane spend altogether?

b) Arif buys food in a shop that costs $8.63. He gives the shopkeeper $10.00. How much change does he get?

c) A shop is selling balls for $6.75 each. How much does it cost to buy 5 balls?

d) John has $18.00. He wants to share it equally between himself and his three sisters. How much money does each person get?

e) A shop is giving 25% discount on toys in a sale. Megan sees a toy that normally costs $40.00. How much does she pay?

f) A bank is giving 12% interest per year on money in the bank. Michael puts in $80.00. How much interest does he get in the year? How much money does he now have in the bank altogether?

Remember!

◆ The money used in a country is called its currency.
◆ There are lots of different currencies used around the world.
◆ Currencies are usually divided into 100 units.
◆ We add, subtract, multiply and divide money in exactly the same way that we add, subtract, multiply and divide decimals.
◆ Percentages are very useful for working out discounts, price increases and interest.

Revision Test on Money

Now that you have worked your way through the chapter, try this revision test. The answers are in the answer book.

1. Joseph and William's mother bought a school backpack for each of them. Each backpack cost $86.00. How much did she pay altogether?

2. A bank pays 15% interest per year on money in the bank. Lucy puts $160.00 in the bank. How much interest does Lucy get in a year?

3. A ticket for a football match normally costs $50.00. The football club decides to give 30% discount off tickets for a match. How much do the tickets cost now? How much do three tickets now cost?

on Money *(continued)*

4. Jerome gets $5.00 pocket money each week. His father decides to increase his pocket money by 20%. How much pocket money will Jerome get each week after the increase?

5. A school teacher bought 10 exercise books for $3.25 each. The shopkeeper then gave her 10% discount. How much did she pay for the 10 books?

Revision Tests

Now that you have worked your way through all the chapters in this book, try these revision tests. The answers are in the answer book.

Revision Test 1

1. 752 = _____ hundreds _____ tens _____ units

2. Write down the value of the underlined digit.

   a) 7̲2   b) 13̲8   c) 1̲067   d) 935̲2   e) 46̲31

3. Put these numbers in order from smallest to biggest.

   1437, 1347, 2298, 1563, 998, 1095

4. Are these true or false?

   a) 4 > 3   b) −4 > −3   c) −1 > 0   d) 2 > −2   e) −3 > −9

5. 837 + 952 =

6. 1046 + 165 =

7. 853 − 341 =

8. 1405 − 86 =

9. There are 36 pupils in class 5 and 29 pupils in class 6. How many more pupils are there in class 5 than class 6?

10. Lucy is 138 cm tall. Gill is 157 cm tall. What is the difference in their heights?

11. 284 × 10 =

12. 364 × 24 =

13. 1242 × 9 =

14. 567 ÷ 7 =

15. A science teacher has 34 test tubes. She keeps the test tubes in test tube racks. Each rack holds 6 test tubes. How many racks does the teacher need?

16. Pineapples are packed into boxes. There are 8 pineapples in each box. At the market there are 13 boxes. How many pineapples are there altogether?

17. What is the lowest common multiple of 8 and 6?

18. What is the highest common factor of 16 and 28?

19. List all the prime numbers between 0 and 20.

20. Complete this table.

21. Write 0.68 as a percentage.

22. Write 23% as a decimal.

Fraction	Decimal	Percentage
	0.3	
$\frac{4}{5}$		

23. Paula got 38 out of 50 in her science test. What was her percentage?

24. In a survey, the ratio of people who liked tea to people who liked coffee was $3:2$. 15 people liked tea. How many liked coffee?

25. What is 15% of 420?

26. Complete these changes of units.

   a) $7\,cm =$ _____ mm   b) $3.4\,m =$ _____ cm   c) $0.83\,kg =$ _____ g

   d) $1230\,g =$ _____ kg   e) $9.5\,l =$ _____ ml   f) $1370\,ml =$ _____ l

27. Susan has $1.2\,l$ of water in a bottle. She drinks $300\,ml$. How much water does she have left in the bottle in litres?

28. So Shan leaves her house to visit a friend at 15:35. She arrives at her friend's house at 16:10. How long did the journey take her?

29. A bus journey took 1 hour and 20 minutes. The journey ended at 11.10 a.m. What time did it start?

30. Say whether these angles are acute angles, right angles, obtuse angles or reflex angles.

   a)   b)   c)   d)

31. What is the size of the angle marked x?

32. Which of these shapes is a regular polygon? What is it called?

   a)   b)   c)   d)

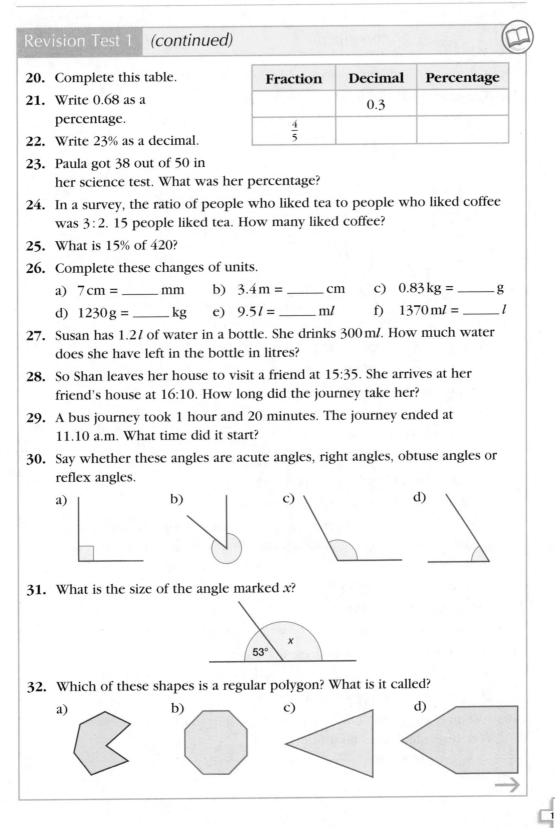

33. A garden is 12 m long and 5 m wide. What is the area of the garden?

34. The edges of a cube are 10 cm long. What is the volume of the cube?

35. 10 people were asked on which day of the week they were born. Here is a bar chart showing the results.

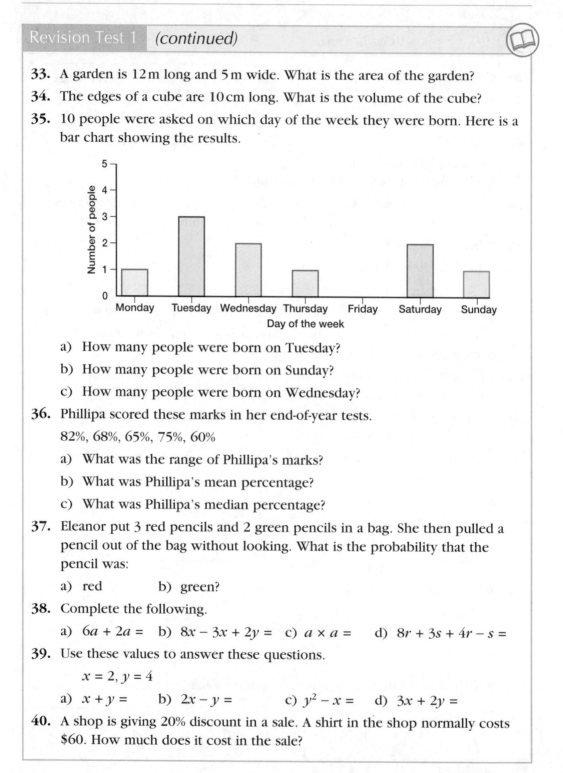

a) How many people were born on Tuesday?

b) How many people were born on Sunday?

c) How many people were born on Wednesday?

36. Phillipa scored these marks in her end-of-year tests.

82%, 68%, 65%, 75%, 60%

a) What was the range of Phillipa's marks?

b) What was Phillipa's mean percentage?

c) What was Phillipa's median percentage?

37. Eleanor put 3 red pencils and 2 green pencils in a bag. She then pulled a pencil out of the bag without looking. What is the probability that the pencil was:

a) red   b) green?

38. Complete the following.

a) $6a + 2a =$ b) $8x - 3x + 2y =$ c) $a \times a =$ d) $8r + 3s + 4r - s =$

39. Use these values to answer these questions.

$x = 2, y = 4$

a) $x + y =$ b) $2x - y =$ c) $y^2 - x =$ d) $3x + 2y =$

40. A shop is giving 20% discount in a sale. A shirt in the shop normally costs $60. How much does it cost in the sale?

1. 2549 = _____ thousands _____ hundreds _____ tens _____ units

2. Put these numbers in order from largest to smallest.

    989, 899, 1063, 1104, 1006, 1603

3. 856 + 143 =

4. 2167 + 353 =

5. 1405 − 236 =

6. Sharmilla is 13 years old. How old will she be in 27 years?

7. There are 212 people in a school. 196 are pupils. 7 are support staff. How many are teachers?

8. 0.8 × 10 =

9. 805 × 42 =

10. 978 ÷ 6 =

11. 17.5 ÷ 10 =

12. 28 CDs are packed into boxes. Each box can hold 5 CDs. How many boxes are needed?

13. Luke is 12 years old. His father is three times older than he is. How old is Luke's father?

14. Which of these numbers are prime numbers?

    9, 11, 13, 15, 17, 19

15. What is the lowest common multiple of 7 and 5?

16. Put these decimals in order from smallest to largest.

    0.3, 0.1, 0.09, 0.25, 1.02, 0.7

17. Put these fractions in order from smallest to largest.

    $\dfrac{7}{8}, \dfrac{3}{4}, \dfrac{1}{2}, \dfrac{4}{2}, 1\dfrac{1}{4}$

18. Write 0.39 as a percentage.

19. $\dfrac{1}{3} \times \dfrac{1}{2} =$

20. $2\dfrac{1}{4} \times \dfrac{1}{2} =$

21. $\dfrac{2}{5} \div \dfrac{3}{5} =$

22. What is 120% of 70?

23. Edward got 62% in his English test. The test was out of 50 marks. How many marks did Edward get?

$\rightarrow$

Revision Test 2 *(continued)*

24. What is 50 m as a fraction of 1 km? Write the answer in the lowest terms.

25. Fill in the gaps in these ratios.

    2 : 3 = 4 : _____ = 8 : _____ = _____ : 15

26. There are 30 children in a class. 18 are girls. What is the ratio of boys to girls in the class?

27. A jug with some water in it has a mass of 2.4 kg. Karin adds 0.5 *l* of water to the jug. What is the mass of the jug and water after Karin adds 0.5 *l*?

28. One day at the North Pole the temperature went up from −9 °C to 4 °C. How much did the temperature go up?

29. John measured the temperature in his classroom. At 8:00 the temperature was 17 °C. By 12:00 the temperature had gone up by 8 °C. What was the temperature at 12:00?

30. Robert is late for a meeting. The meeting starts at 10.45 a.m. Robert arrives 25 minutes late. What time does he arrive?

31. 20% of Emma's mass is 8 kg. What is Emma's mass?

32. What is the size of the angle marked x?

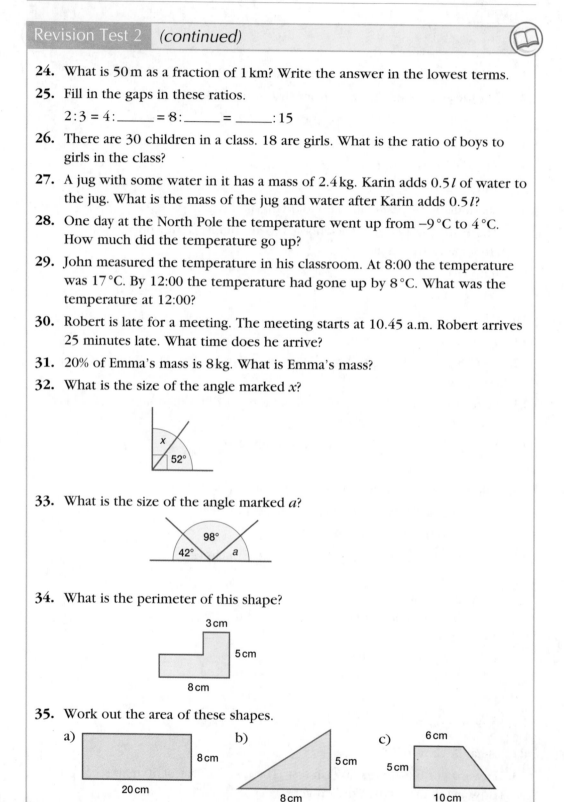

33. What is the size of the angle marked a?

34. What is the perimeter of this shape?

35. Work out the area of these shapes.

36. What is the volume of this solid shape?

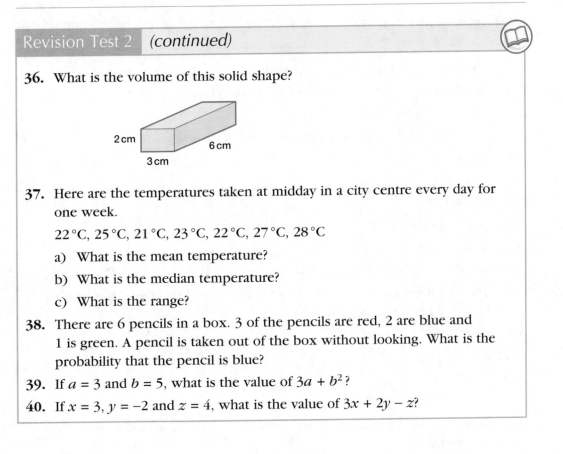

2 cm
3 cm
6 cm

37. Here are the temperatures taken at midday in a city centre every day for one week.

22 °C, 25 °C, 21 °C, 23 °C, 22 °C, 27 °C, 28 °C

a) What is the mean temperature?

b) What is the median temperature?

c) What is the range?

38. There are 6 pencils in a box. 3 of the pencils are red, 2 are blue and 1 is green. A pencil is taken out of the box without looking. What is the probability that the pencil is blue?

39. If $a = 3$ and $b = 5$, what is the value of $3a + b^2$?

40. If $x = 3$, $y = -2$ and $z = 4$, what is the value of $3x + 2y - z$?

Revision Test 3

1. 4605 = _____ thousands _____ hundreds _____ tens _____ units

2. Put these numbers in order from smallest to largest.

   1003, 989, 990, 1100, 1253, 1000

3. Put these numbers in order from smallest to largest.

   −6, 4, −1, 0, 12, −4, −12, 1

4. 936 + 63 =

5. 487 + 134 =

6. 1895 − 184 =

7. 3006 − 328 =

8. Saul is 168 cm tall. Thomas is 13 cm taller than Saul. How tall is Thomas in cm?

9. Two numbers added together give an answer of 357. One of the numbers is 219. What is the other number?

10. 0.03 × 100 =

11. 248 × 14 =

12. 186.7 ÷ 10 =

13. 903 ÷ 15 =

14. A café uses 7 l of milk every day. How many litres of milk does the café use in 3 weeks?

15. Manjit has a piece of rope that is 14 m long. Manjit divides the rope into 4 equal pieces. How long is each piece in cm?

16. What is the lowest common multiple and the highest common factor of these three numbers?

    6, 8, 12

17. Complete this table.

Fraction	Decimal	Percentage
		25%
$\frac{3}{4}$		

18. Are the following true or false?

    a) $\frac{3}{4} > \frac{1}{2}$

    b) $\frac{1}{4} > \frac{1}{3}$

    c) $\frac{7}{8} < \frac{19}{24}$

    d) $\frac{9}{7} = 1\frac{2}{7}$

19. $\dfrac{1}{6} + \dfrac{2}{3} =$

20. $\dfrac{7}{8} - \dfrac{3}{4} =$

21. $\dfrac{4}{5} \times \dfrac{3}{7} =$

22. $\dfrac{6}{7} \div \dfrac{2}{3} =$

23. There are 190 children in a school. 20% of the children are in the school choir. How many children are in the school choir?

24. Femi goes to the shops and spends $\frac{1}{4}$ of his money on bread and $\frac{3}{8}$ on meat. What fraction of his money has Femi spent? What fraction of his money does he have left?

25. Round these decimals to 2 decimal places.

    a) 4.148   b) 5.092

    c) 0.458   d) 17.315

26. Jim has a mass of 42 kg. His mass is 10 times bigger than the mass of his baby brother. What is the mass of Jim's baby brother in grams?

27. A bus travels for 45 km from one town to another. It then travels another 7500 m to the next village. How far has the bus travelled altogether in km?

28. A school sports day started at 14:30 and lasted for 3 hours and 25 minutes. What time did it finish?

29. An airplane took off at 09:30 and landed at 15:10. How long did the flight take?

30. If the temperature starts at −7 °C and goes up by 15 °C, what is the new temperature?

31. What is the size of the angle marked *x*?

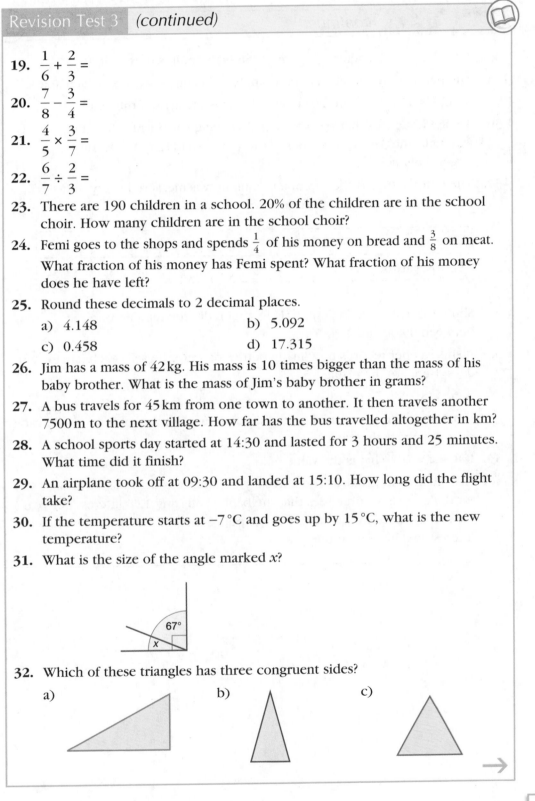

32. Which of these triangles has three congruent sides?

    a)   b)   c)

33. A circle has a diameter of 17 cm. What is the radius of the circle?

34. The perimeter of an equilateral triangle is 24 cm. How long is each side?

35. A field is 40 m long and 20 m wide. What is the area of the field?

36. A cube has edges that are 4 m long. A rectangular prism is 5 m long, 4 m wide and 3 m high. Which of these two solid shapes has the largest volume?

37. The temperature at the top of a mountain was measured every hour between 07:00 and 12:00. Here are the results.

Time	07:00	08:00	09:00	10:00	11:00	12:00
Temperature	0 °C	2 °C	5 °C	9 °C	12 °C	14 °C

Show this data as a line graph. How much did the temperature rise between 09:00 and 11:00?

38. A shopkeeper records how many bottles of water he sells each day for ten days. Here are the results.

8, 6, 4, 8, 3, 4, 7, 4, 5, 1

What were the mean, median and mode of the number of bottles sold each day? What was the range?

39. If $8 + x = 20$, what is the value of x?

40. George lives in Jamaica. The currency in Jamaica is the dollar. The sign for dollars is $. George sees that the bank is offering 12% interest per year for money saved in the bank. George puts $3000 in the bank. How much interest will he earn in one year?